HAPM Guide to Defect Avoidance

HAPM Guide to Defect Avoidance

Christopher Garrand

Construction Audit Ltd.

London and New York

First published 2001
by Spon Press
11 New Fetter Lane, London EC4P 4EE

Simultaneously published in the USA and Canada
by Spon Press
29 West 35th Street, New York, NY 10001

Spon Press is an imprint of the Taylor & Francis Group

Designed and typeset by Paul Jones Associates
Printed and bound in Great Britain by Butler and Tanner Ltd, Frome and London

The publisher makes no representation, express or implied, with regard to the accuracy of the
information contained in this book and cannot accept any legal responsibility or liability for
any errors or omissions that may be made.

British Library Cataloguing in Publication Data
A catalogue record for this book is available from the British Library

Library of Congress Cataloguing in Publication Data
A catalogue record for this book has been requested

ISBN 0-419-24890-0

Author's Note:
This guide was produced by Construction Audit Ltd. (CAL), a subsidiary of Building
Performance Group Ltd. It was conceived and written by Christopher Garrand, drawing on
the 10 years auditing experience of CAL's professional staff – the contributions of Will
Johns, Julian King, Diane McGlynn, Brenda Major and Paul Sharp are specifically
acknowledged.

All technical enquiries should be addressed to Construction Audit Ltd. at
International House, 26 Creechurch Lane, London EC3A 5BL.
Telephone: 020 7929 2889 Facsimile: 020 7929 1366.

The author makes no representation, express or implied, with regard to the accuracy of the
information contained in this book and cannot accept any legal responsibility or liability for
any errors or omissions that may be made.

Contents

Foreword

Work carried out by the BRE on quality on new build construction — much of which was done in conjunction with professional staff now at Construction Audit Ltd. — revealed that responsibility for defects was fairly equally divided between site and design office, with materials and components forming a relatively small proportion of the total. Workmanship aside, it is how designers use materials and components that causes problems.

More recently, the Egan report has drawn the attention of the building industry to the need for improvements in efficiency, and in particular the need to reduce the number of defects which occur during construction work. This *HAPM Guide to Defect Avoidance* should make a significant contribution to these aims, reminding designers of their responsibilities and identifying the crucial defects which, through error or oversight, can have unfortunate consequences for building owners.

Criteria cited in the document are both qualitative and quantitative, though it is important to appreciate that, in the case of the former, professional judgement and — frequently — experience is needed to ensure defect-free design. It must also be appreciated that, while it may be relatively straightforward to meet performance criteria for separate parts of the total design, it is in the assembly of these partial solutions that lack of co-ordination is often most apparent — when specifying according to performance, it may still be essential for designers to constrain the range of solutions to ensure that delegated parts of the design will still fit together satisfactorily on site; it is no use relying on blanket requirements that all work is to be in accordance with building regulations or all relevant codes, when data indicates that substantial numbers of defects (25% is by no means untypical) result from non-compliance with even these most basic of requirements.

Design is rarely entirely right or entirely wrong, and it is often only under exceptional circumstances that problems or deficiencies are revealed. This guidance should help designers to identify the pinch points in their designs. If some of the points appear to be of an elementary nature, then that is simply a reflection of what is found in practice.

It is worth noting a few examples from my own experience of the kinds of things that still cause problems in housing. Many designers do not seem to understand the need for sub-floor ventilation, let alone the need to check the free area of different types of airbrick. Shrinkage and curling of screeds is another item rarely appreciated, as is rainwater discharge at valleys overshooting the gutters in periods of heavy rain, and the thermal movement of plastics components. When it comes to adequacy of information made available to site staff, perhaps one of the more serious of items is the question of lateral restraint between walls and floors and walls and roofs. Strapping can become quite complex in certain circumstances, and clarity of information is vital if the site is to get it right.

However long the formal period of training of designers there are bound to be gaps in their knowledge, particularly where the pace of change in the introduction of new materials and techniques is accelerating — they need all the help they can get.

HAPM are to be congratulated on the quality of this essential aide-mémoire for the designer.

H W (Harry) Harrison, ISO Dip Arch RIBA
formerly Head of Construction Practice Division, Building Research Establishment

May 2000

Introduction

The *HAPM Guide to Defect Avoidance* has been produced by Construction Audit Ltd. in their role as technical auditors to Housing Association Property Mutual (HAPM), the insurance company set up by a group of housing associations in 1990 to provide protection against the consequences of any unforeseen defects in the construction of their developments. The *Guide* builds on its predecessor, the *HAPM Defect Avoidance Manual* (published 1991), drawing on the extensive knowledge of building defects acquired by the HAPM technical audit.

The technical audit is fundamental to the HAPM insurance cover. Utilising the techniques of risk management, it seeks to identify and — as far as possible — eliminate defects which may give rise to a claim under the HAPM policy. It is a staged, iterative process.

The first stage involves an appraisal by the technical auditors of the available scheme documentation (e.g. drawings and specifications), the results of which are forwarded to the housing association (or social landlord) and their consultants or contractors in the form of a report. The report highlights any areas of uncertainty, ambiguity or concern. An auditor then visits the scheme at critical stages during construction, reassesses the design in the light of what is found on site and forms an opinion as to the general quality of workmanship. The audit is then updated, taking account of both the site auditor's findings and any new information received by HAPM, including the responses of consultants and contractors to the original audit. Further site visits may be undertaken and the process of updating the audit repeated if the scheme is large or complex or gives any cause for serious concern.

It must be stressed that the technical audit is not intended as a full check on every aspect of every part of a scheme. It is a mechanism for ensuring that accepted and defined standards of good practice are understood and implemented. HAPM does not strive to set any new standards, nor does it seek to impose any specific design solution or method of working. Similarly this *Guide* aims to promote good practice, not to act as a comprehensive textbook for building construction.

Nevertheless, Construction Audit Ltd, informed by HAPM's first hand experience in dealing with claims resulting from poor design, has brought together in a single volume a wide range of advice on the design of low-rise, newbuild housing. It is intended to complement — not replace — the sometimes bewildering array of information available from bodies such as the British Standards Institution, the Building Research Establishment and trade organisations like the Brick Development Association and TRADA. The aim is to promote a better understanding of:–

• How defects in design may manifest themselves in a range of different building elements.

• The obligation on designers to make informed technical decisions that take due account of the vast amount of published guidance available to the building industry.

• Why — in construction terms — 'best practice' is always preferable to the 'bare' minimum.

The *Guide* is divided into six chapters:–

▓ FOUNDATIONS AND GROUND FLOOR STRUCTURES
▓ EXTERNAL WALLS
▓ ROOFS
■ INTERNAL WALLS AND FLOORS
▓ ABOVE GROUND SERVICES
▓ BELOW GROUND DRAINAGE AND EXTERNAL WORKS

Underlying the content and structure of the *Guide* is the belief that defects in design can only be fully understood in terms of their consequences, and that a clear distinction must be drawn between the two.

This crucial distinction between cause and effect reflects the fundamental difference between the immediate priorities of building owners and occupiers, and the long-term views of those seeking to avoid or rectify defects. The former are often only concerned that a roof leaks, not what is causing it to leak, sometimes leading to attempts at rectification which are either ineffective or give rise to further problems. The latter are primarily interested in ascertaining *why* a roof should leak, so that they can ensure that the problem is avoided in the first place.

Each chapter of the *Guide* begins with a double-page spread that links a list of design defects to a short, illustrated description of their potential consequences. The defects have been selected and 'rated' on the basis of experience gained in auditing over 4,000 newbuild housing schemes (over 75,000 units in all) and dealing with an ever increasing number of claims arising from defective design. The rating, on a scale of 1 to 4, indicates the potential risk inherent in each defect, as perceived by HAPM (i.e. the probability of the designer 'getting it wrong' combined with the likely severity of its consequences).

Every defect is then presented in the context of a 'problem' (an explanation of why it is necessary for designers to consider the issues), followed by guidance on how to avoid it. Common design mistakes are then highlighted. Finally, a list of references directs the reader to more detailed guidance, with specific clause and page numbers where appropriate.

The *Guide* ends with a comprehensive bibliography, up to date as of April 2000. However, changes and revisions are inevitable, especially with the increasingly rapid transition to European Standards. Thus the most recent version of any document cited should always be consulted. In order that the *Guide* should have the broadest application, no specific reference is made to any documents issued in support of national building regulations. It should also be noted that, unlike the *Defect Avoidance Manual,* the *Guide* focuses only on design defects. Since 1992, materials and components have been dealt with in the *HAPM Component Life Manual,* and defects in workmanship are now covered by the *HAPM Workmanship Checklists* (published 1999).

When HAPM was formed, the technical audit was an innovative approach to the management of building defects in social housing. A decade on and the audit process is still being developed and refined in response to a continuous feedback of data.

Though based on the specific concerns of an insurer, it is hoped that the *HAPM Guide to Defect Avoidance* will be of value to all those involved in the design and management of low-rise housing of traditional construction. Drawing on HAPM's unique insight into the UK building industry, its presentation of much tried-and-tested information in a new way should be of benefit to all those who strive for better and more reliable building.

Abbreviations and Acronyms

app.	appendix
ch.	chapter
chs.	chapters
cl.	clause
cls.	clauses
fig.	figure
figs.	figures
m	metres
mm	millimetres
m^2	square metres
mm^2	square millimetres
m^3	cubic metres
mm^3	cubic millimetres
p.	page
pp.	pages
tab.	table
vol.	volume
BCA	British Cement Association
BDA	Brick Development Association
BPF	British Plastics Federation
BRE	Building Research Establishment
BS	British Standard
BSI	British Standards Institute
BSRIA	British Services Research and Information Association
CDA	Copper Development Association
CIRIA	Construction Industry Research and Information Association
CP	Code of Practice
CPDA	Clay Pipe Development Association
DAS	Defect Action Sheet
DD	Draft for Development
DG	Design Guide
DMRB	Design Manual for Roads and Bridges
DN	Design Note
EN	European Standard
FIDOR	Fibre Building Board Organisation
FS	Fact Sheet
GBG	Good Building Guide
GGF	Glass and Glazing Federation
GPN	Good Practice Note
HA	Highways Agency
HAPM	Housing Association Property Mutual
ICRCL	Interdepartmental Committee for the Redevelopment of Contaminated Land
IOP	Institute of Plumbing
IP	Information Paper
LSA	Lead Sheet Association
MAC	Mastic Asphalt Council
NHBC	National House Building Council
NRPB	National Radiological Protection Board
NSAI	National Standards Authority of Ireland
PESDG	Plumbing Engineering Services Design Guide
PIFA	Packaging and Industrial Films Association
SP	Special Publication
SR	Standard Recommendation
SWA	Steel Windows Association
TN	Technical Note
TRA	Trussed Rafter Association
TRADA	Timber Research & Development Association
WIS	Wood Information Sheet
WPIF	Wood Panel Industries Federation
ZDA	Zinc Development Association

Foundations and Ground Floor Structures

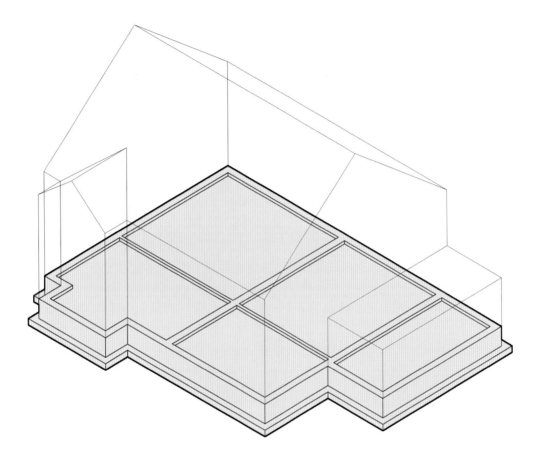

Defects and their Consequences

DEFECT	CONSEQUENCES			
	A	B	C	D
INCOMPLETE OR POORLY FOCUSED SITE INVESTIGATIONS	3	3	2	2
FOUNDATIONS NOT SUITED TO GROUND CONDITIONS	1	1		
FAILURE TO DESIGN FOR DIFFERENTIAL MOVEMENT	4	3	1	
INSUFFICIENT DEPTH OF FOUNDATIONS IN CLAY SOILS	4	3		
INADEQUATE SPECIFICATION OF PILED FOUNDATIONS	2	2		
INADEQUATE SPECIFICATION OF VIBRATORY GROUND IMPROVEMENT	3	3		
CONCRETE FLOORS NOT PROTECTED AGAINST HEAVE	2	4	2	
UNEVENLY SUPPORTED GROUND BEARING FLOORS		2	1	
GROUND FLOORS NOT PROTECTED FROM DAMP			2	
POORLY VENTILATED SUB-FLOOR VOIDS			3	
INADEQUATE PRECAUTIONS AGAINST GASEOUS CONTAMINANTS				4
INADEQUATE PRECAUTIONS AGAINST NON-GASEOUS CONTAMINANTS				3

The four columns labelled A to D refer to the four types of consequence detailed on the opposite page.

KEY TO HAPM RATING

1 Low probability of defect occurring, and only likely to have minor consequences.

2 Low probability of defect occurring, though with potentially serious consequences.
or
Reasonable probability of defect occurring, though only likely to have minor consequences.

3 Reasonable probability of defect occurring and with potentially serious consequences.
or
High probability of defect occurring, though only likely to have minor consequences.

4 High probability of defect occurring and with potentially serious consequences.

Consequences

A DISTORTION AND CRACKING OF STRUCTURE

Defects in the design of foundations — and sometimes ground floors — will invariably result in movement and hence in the distortion or cracking of any structures they support. Cracking due to structural movement appears in the horizontal and vertical planes of mortar joints, and often runs at oblique angles; cracks are also likely to vary in width, indicating the direction of movement (cracks due to thermal movement are generally of uniform width and run straight through materials regardless of joints). Leaning or bulging external walls or other supporting structures, and cracks at the junctions of walls and ceilings are often the result of structural movement too. Likewise the disruption of claddings and wall finishes, and the displacement of flashings and other details remote from the ground.

B DISTORTION AND CRACKING OF GROUND FLOORS

Uneven and cracked floor surfaces, the disruption of floor finishes and cracking at floor junctions are all indicators of structural movement, which can be as much the result of defects in the design of the floors themselves as an indication of deficiencies in the adjacent foundations.

C WATER STAINING, MOULD GROWTH AND FUNGAL DECAY

Moisture is always potentially harmful to buildings. Water penetration via ground floors can stain, discolour and cause the deterioration of floor finishes. Condensation will encourage mould, and any moisture can lead to timber or timber-based components within or adjacent to the floor being attacked by fungal decay (identification can be aided by the BRE's *Recognising wood rot and insect damage in buildings*). Bear in mind that moisture-related damage relating to ground floors is just as likely to be caused by disruptions to damp proof membranes or door thresholds (i.e. a consequence of structural movement) as it is by defects in the design of damp-proofing.

D DIRECT THREAT TO HEALTH

Gaseous and non-gaseous contaminants can be present in many types of ground. Some, like radon, arise naturally. Many others, including heavy metals and toxic chemicals, are the result of human activity; the present or former use of the site may well indicate if the soil is likely to be contaminated. There are also contaminants that can be natural and man-made. Sulfate is a good example. The danger with many contaminants is that they can be inhaled or ingested (either directly or indirectly via the food chain), although some harmful contaminants can cause problems by skin contact alone. Any failure to identify and deal with suspected contaminants within the ground can result in the occupiers of buildings and sites being exposed to potentially serious threats to their health.

Incomplete or Poorly Focused Site Investigations

A staged, methodical site investigation – appropriate to the scale of the development – is a prerequisite to the design of any building, especially its foundations, which must be matched to the ground conditions, existing and anticipated. However, site investigations are often incomplete or poorly focused, and it is not always appreciated that a 'ground' investigation is – in itself – not enough; the site must be seen as a whole if problems are to be anticipated and avoided.

GUIDANCE

The objectives of any site investigation must be decided at the outset, since they will determine its nature. The investigation should progress through the following stages:

- **Desk study**: At the very least this should entail a study of old maps, geological and mining information, the records of local and statutory authorities, and any previous site investigations. Aerial photographs, meteorological data, and information held by the environment or river authorities may – depending on the project – also be of use.

- **Site reconnaissance**: This involves a visual assessment of the site and its locality (i.e. a 'walk over' survey). Desk-study information is checked and landscape features that may influence ground conditions (e.g. nearby rivers or landfill sites) are recorded. Nearby cuttings and quarries can help in assessing the likely ground conditions; damage to buildings on adjacent sites may indicate a settlement problem. Older properties may have been sited to avoid poor ground or flooding; and the value of verbal information – especially from older persons – should not be underestimated.

- **Detailed examination**: It is at this stage that the ground investigation is carried out. The positioning of trial pits and bore holes should enable the nature of the soil, its bearing capacity, shrinkability (if clay), and sulfate content to be fully described. Ground water levels should be established, and unstable or contaminated ground identified. Laboratory testing or on-going monitoring may be required.

- **Follow up investigations**: The findings of the ground investigation should be reviewed as and when the ground is disturbed by construction operations.

Although stages may overlap, the importance of the desk study and site reconnaissance being carried out prior to the ground investigation cannot be stressed too highly. The failure to carry out such an exercise makes it very difficult to ensure that the nature of any ground investigation is proportional to the scale of the project, and that design decisions are not based on unrepresentative or misleading information.

Spoil from trial pit shows different soils

COMMON MISTAKES

- No desk study.
- No site reconnaissance.
- Failure to grasp distinction between site and ground investigations.
- Too few or mis-positioned trial pits or bore holes.
- Inadequate investigation of ground contamination.
- Investigations not reviewed during construction.

REFERENCES

BS 5930:1981
BS 8004:1986 cl.2.2.1
BS 8103–1:1995 cls.6.1–6.3
BRE Digest 64, 348 & 427–2
NHBC ch.4.1

Foundations not Suited to Ground Conditions

THE PROBLEM   RATING 1

Foundations ensure that the ground on which a building stands does not — when loaded — deform or settle so as to cause potentially damaging structural movement. Although "A building, its foundation and the supporting soil interact in a complex manner, the behaviour of one depending upon, and influencing, that of the others" (BRE Digest 63), the generally light loading of low-rise housing tends to mean that it is ground conditions which primarily guide the selection of the foundation system. Designers must, when considering the findings of the ground investigation, appreciate which types of foundation are best suited to which types of ground.

GUIDANCE

Shallow strip or pad foundations are suited to:

- Dense sands or gravels with a water table below foundations level, or where a high water table can be controlled (e.g. by pumping).
- Firm or hard clays, although depths may have to be increased to account for seasonal variations in the moisture content of the ground.
- Rock [near ground level], notwithstanding that there may be problems if the rock is thinly bedded or heavily shattered, or where soft shales, mudstone or chalk are present.

Stable ground may permit the use of trench fill foundations, while reinforced wide strips may be required on weak, though uniform ground. Pads are useful for distributing concentrated loads. Strips or pads may also be appropriate on loose granular soils which have been deeply compacted, or where poor, soft ground has been excavated and replaced.

Short ('mini') piles may be appropriate where a good bearing exists within 2 to 4 metres of finished ground level and where the upper strata are weak (e.g. made ground), the ground is of shrinkable clay or the water table is high.

Rafts can be used on soft natural ground, or where ground conditions are uneven or where mining subsidence, excessive shrinkage or swelling are likely to be a problem.

Deep foundations (e.g. piles) are generally required where the ground comprises soft clays, peat, organic soils, or loose sands or gravels near ground level, or where there are voids within the ground (e.g. old mine workings). The latter may dictate grouting or some other form of void-filling. If the ground is unsuited to piles or if the bearing strata is very deep, advanced engineering solutions such as caissons may have to be employed.

Excavation for shallow strip footings

COMMON MISTAKES

- Foundation design does not account for high level of water table.
- Trench fill footings used in unstable ground.
- Strip footings used where short piles may be more appropriate.
- Insufficient allowance for void-filling (grouting).

REFERENCES

BS 8004:1986 cls.3 & 4
BS 8103–1:1995 cls.7.2 & 7.3
BRE Digests 63, 64 & 67

Failure to Design for Differential Movement

Most new buildings will, to some degree, move downward ('settle') as the ground compresses under the pressure of their foundations, depending on the weight of the building and the nature of the subsoil; a lightweight entrance porch on stiff, compact gravel will settle significantly less than a heavy, loadbearing masonry wall on soft silty clay. Mining operations or other activity within the ground can also cause downward movement ('subsidence'), while upward movement ('heave') can result from the swelling of the ground following the removal of trees in clay soils or reduction in ground loading (e.g. after excavation for a basement or the demolition of a heavy structure). While settlement, subsidence or heave are often unavoidable, damage can be avoided if all parts of a building are designed to move together, in the same direction and by the same amount. Foundations and ground works must be designed to accommodate variations in applied load and uneven ground conditions if 'differential' movement is to be avoided.

GUIDANCE

Variations or concentrations in the load applied to foundations — e.g. the result of abrupt changes in the height of walls, or large openings near ground level — can be accommodated by either designing the entire foundation to suit the greatest loads or by providing localised reinforcement. Raft foundations can be locally thickened in lieu. Particular care should be taken to ensure that lightweight structures, such as bays or porches, are designed to move in exactly the same way as any heavier structure to which they are attached. They should share the same foundations or be designed to move independently. Movement joints in the foundations may be necessary.

The simplest way of dealing with uneven ground conditions — which can cause settlement, subsidence or heave — is to utilise foundations which avoid the problem altogether. Piles can be used to take the load of the building down to a uniform bearing, while a system of beams and pads or a properly engineered raft can distribute loads in a way which either 'bridges' soft-spots (e.g. where old excavations have been filled with inadequately compacted material) or which reduces the stresses on the weakest ground. These techniques are especially suited to situations where subsidence or heave may occur. Obstructions within the ground (e.g. old basements) can be dealt with in the same way, although it may be easier to remove them altogether or to incorporate old footings within a new system.

Shallow footings may also be used in conjunction with ground improvement techniques, ranging from localised compaction to complex vibro-displacement or replacement techniques.

Shallow pad footing for post (walls piled)

COMMON MISTAKES

• No provision for movement of porches, bin stores or similar attached structures.

• Foundation design does not accommodate soft spots or obstructions in the ground.

• Movement joints provided in foundations but not aligned or continuous with those in structure above.

REFERENCES

BS 8004:1986 cl.2.3.2
NHBC ch.4.4–D4(b)

Insufficient Depth of Foundations in Clay Soils

THE PROBLEM   RATING 3 – 4

Clay minerals swell as they take up moisture and shrink as they dry. This means that clay soils are constantly heaving and subsiding in response to seasonal changes in the weather, often by amounts that are large enough to cause significant damage to buildings sitting on strip or pad foundations. But the influence of the weather on the soil decreases with depth, meaning that foundations in clay soils will be less prone to damage the deeper they are. However, the depth at which clay soil can be considered 'stable' depends on a number of factors, including the degree to which it can take up and lose moisture (its 'shrinkage potential') and the water demand of any nearby trees; both are crucial in determining the depth of a simple foundation in clay soil.

GUIDANCE

The shrinkage (or 'heave') potential of a clay soil is described by its 'plasticity' index. Soils with an index of 40% or more are classed as high shrinkage, those below 20% as low, those between 20% and 40% as medium. The water demand of a tree relates to its species, as does its likely mature height.

No trees in proximity: if the foundations are outside of the zone of influence of any trees (0.5, 0.75 or 1.25 times the mature height for low, moderate and high water demand respectively), then their minimum depth should be 1.0m for high shrinkage soil, 0.9m for medium, or 0.75m for low.

New trees in proximity: if it can be guaranteed that no trees will ever be planted closer to the foundations than 0.2, 0.5 or 1.0 times their mature height (low, moderate and high water demand respectively), then the minimum depth of foundation should be 1.5m for high shrinkage soil, 1.25m for medium and 1.0m for low. For trees closer than these distances, foundation depth should be derived from tables published by the NHBC.

Existing trees in proximity: use NHBC tables.

Existing trees removed: use NHBC tables, with depth measured from original ground level; the height of the tree at the time of removal may be used in lieu of its mature height.

Strip footings deeper than 1.2m are not recommended. Trench fill deeper than 2.5m should be properly engineered, although at this depth, ground beams and short piles may be more economic.

Foundations deeper than 1.5m within the zone of influence of existing trees (even if removed) are particularly vulnerable to the horizontal forces associated with heave. Where soils have a high or medium heave potential, compressible material or void formers should be provided against the inner faces of all footings to external walls.

Use of compressible layers

COMMON MISTAKES

- Foundations not designed to accommodate future tree planting.
- No account taken of trees beyond site boundary.
- Failure to identify species (hence water demand and likely mature height) of existing trees.

REFERENCES

BS 8103–1:1995 cl.7.3.8
BRE DAS 96
BRE Digests 240, 241, 242 & 298
NHBC chs.4.2 & 4.4–D8

Inadequate Specification of Piled Foundations

THE PROBLEM   RATING 2

Although the detailed design of piled foundations is invariably carried out by a specialist contractor, the outline design remains the responsibility of the building designer, usually a structural engineer. They produce the piling layout, design the ground beams or pile caps and draft a 'performance' specification, and hence must understand the principles of pile design and the problems that may be encountered.

GUIDANCE

The main types of pile used in domestic construction are cast in situ concrete piles and precast concrete driven piles.

Piles derive their bearing capacity from a combination of the friction along their sides (i.e. between the ground and the pile) and the end bearing of their base. In broad terms, bored piles rely on friction, driven piles on end bearing.

The spacing of piles — which may be circular or square in section — is determined by the nature of the ground, the behaviour of the piles in groups and whether they are bored or driven. However, as a general rule:

- **Friction (bored) piles** should be spaced at not less than three times the length of their perimeters, centre to centre, which for circular piles is just over three times their diameter.

- **End bearing (driven) piles** should be spaced so the distance between adjacent shafts is no less than the least width of the piles, although special considerations will apply where piles have enlarged 'undercut' bases.

The tops of piles should be connected by ground beams or pile caps so that piles cannot rotate, twist or move laterally (i.e. pile heads must be restrained at 90° in at least two directions).

Other points that need to be taken into account include:—

- The possibility of 'down drag' (negative skin friction) due to the consolidation of compressible strata (e.g. silt or peat) — the allowable loads on the piles may have to be decreased, or the piles coated or sleeved to reduce the friction.

- The effects of clay heave, especially where trees or hedges have been removed — the upper parts of the piles may have to be sleeved or the piles reinforced, and ground beams designed to incorporate compressible layers.

Requirements for testing and recording must be explicit. Load tests (not usually necessary for friction piles which have a factor of safety between 2 and 3) determine the final settlement of a pile, its ultimate bearing capacity or its soundness. Electronic tests can check the integrity of concrete. Pile diameters and installed lengths should be recorded daily.

Pile caps ready to receive ground beams

COMMON MISTAKES

- Piles too close.

- No allowance for negative skin friction or clay heave.

- Ground beams not set above centre lines of piles.

- Single piles (e.g. at isolated columns) not tied in at least two directions.

- No requirements for testing included in the specification.

REFERENCES

BS 8004:1996 cl.7
BRE Digests 241 & 242
NHBC ch.4.5

Inadequate Specification of Vibratory Ground Improvement

THE PROBLEM   RATING **3**

Vibratory ground improvement techniques are used to strengthen weak, loose or compressible soils, usually to enable the use of shallow strip or raft foundations in lieu of deep piling. Large vibrating tubular 'pokers' are inserted into the ground, the aim being either to compact the ground or to reinforce it with columns of stronger material. Whilst the detailed design of such work is usually undertaken by a specialist contractor, responsibility for determining – amongst other things – the loads to be imposed on the foundations and the acceptable degree of variation between the predicted and actual characteristics of the improved ground (checked by post-treatment testing) rests with the designers of the building, who must understand the principles, methods and limitations involved.

GUIDANCE

Deep compaction: Suitable for inherently strong but loose, non-cohesive granular soils (e.g. sands and gravels) or fill. The poker is allowed to sink under its own weight, displacing the non-cohesive soil and forming a shallow crater at ground level, which is filled with sand or gravel as the compaction proceeds (medium to coarse gravel should be used if the ground contains layers of cohesive or low permeability soil). This results in highly compacted areas of ground between 2m and 4m in diameter. Groups or rows of compactions are placed to suit the plan of the foundations, usually simple strips.

Strengthening with 'stone' columns: Suitable for soft silts and clays or weak compressible fills; can also be used instead of deep compaction. The poker is used to form columns of coarse granular material (i.e. clean, hard and inert stone) by one of two methods:–

- **Vibro-displacement:** Ground which is strong enough to support itself is *displaced* by the vibrator (i.e. the ground between the columns is subject to some compaction), which is then removed to allow the stone to be tipped and compacted in 1m layers. Also known as the 'dry' process method.

- **Vibro-replacement:** Water is jetted down through the end of the poker then upward to form a surrounding void. The water supports the unstable sides of the bore while the stone backfill is placed, prior to compaction by the poker as it is withdrawn. Also known as the 'wet' process method.

Columns are spaced at a maximum of 2m intervals along the centre lines and at each intersection of reinforced concrete strip foundations, which should be at least 600mm deep and designed to span the columns. Raft foundations can also be used.

Vibro displacement or replacement techniques are not suitable for stiff or very soft clays, or ground with peat layers close to foundation level. Ground containing soils subject to collapse or rising water levels, filled ground including voids or which may be subject to further settlement or which includes 15% or more of degradable organic material, and highly contaminated ground are also unsuitable.

Head of a vibro stone column

COMMON MISTAKES

- Responsibilities of engineer and specialist contractor not defined at the outset.
- No account taken of effect of treatment on water table.
- Excavations or alteration of site levels permitted after treatment of ground.
- Soakaways positioned in proximity to treated ground.
- Limestone backfill used in acidic ground.
- No requirements for testing included in the specification.

REFERENCES

BS 8004:1986 cls.6.6.3 & 6.6.4

BRE Digest 427–2

BRE IP 05/89

BRE Report 391

NHBC ch.4.6

Concrete Floors not Protected against Heave

Although the ground beneath most buildings is relatively well protected against the sort of climatic variations which cause the shrinkage or swelling ('heave') of clay soils, it is not entirely immune from such movement. Trees or shrubs in close proximity may cause seasonal fluctuations in the moisture content of the ground, as can nearby water courses or land drainage systems. But the greatest threat to ground floors is the heave that takes place after the felling or removal of trees and hedgerows, or where tree roots have been severed by excavation. This is because trees or shrubs often take more water out of the ground than can be replenished over the winter months, leaving the natural ground in a permanently dehydrated and shrunken ('desiccated') state. The removal of the trees and shrubs removes the demand for water, resulting in the rehydration of the ground and the sort of swelling and upward pressure that can cause cracking and displacement. Hence the need to protect concrete floors from the effects of heave in clay soils.

GUIDANCE

The most reliable way of protecting a concrete ground floor from the possible effects of heave is to use a suspended form of construction (e.g. precast planks or beam-and-block), with the void below being deep enough to keep the ground from touching the underside of the structure, even when the ground has swollen to its full extent. When deciding a suitable depth of void, the heave (shrinkage) potential of the soil and the need to keep clear any ventilation paths must be considered. The following will normally suffice:-

Heave Potential of soil	Unventilated Void	Ventilated Void
High	150mm	225mm
Medium	100mm	175mm
Low	50mm	125mm

If the heave potential of the soil in unknown, it should be assumed to be 'high'. For beam-and-block floors, the void depth should be measured from the underside of the beams, not the infill blocks.

Reinforced in situ floors may also be used if laid on a suitable void former, which must be of a type which is designed to support – as formwork – the wet concrete yet still limit the pressure exerted on the underside of the slab by the expanding clay soils. This generally means void formers which are of a cellular or 'egg crate' pattern. Low density polystyrene slabs, as used to protect ground beams or the sides of footings, are not suitable. The thickness of any void former should be the same as the depth of an unventilated void.

Under certain circumstances, heave may be countered by a reinforced, fully engineered raft.

Void beneath a concrete ground floor

COMMON MISTAKES

- Floor voids not deep enough to accommodate ventilation.

- Depth of void beneath beam and block floors measured from underside of blocks, not beams.

- Polystyrene slabs used as void formers.

REFERENCES

BS 8103–4:1995 cl.5.3

BRE Digest 298

NHBC chs.4.2–app.E & 5.2–D10

Unevenly Supported Ground Bearing Floors

THE PROBLEM   RATING 1 – 2

Simple ground bearing slabs — usually of in situ concrete — are effectively short, fat columns that distribute load across a large area of ground. They are not intended to act as beams, and it is therefore important they are laid on a level, even base, since variations in the loading of the ground can result in the slab being forced to 'span' or 'cantilever' in ways that were not anticipated. However, variations of the soil means that it is rarely possible to cast a slab direct on the ground. Compressible clays may be mixed with sands or gravels creating 'soft' and 'hard' spots. Outcrops of rock, retained structures within the ground and old excavations can also cause problems, even over a relatively small area. Hence the necessity for the floor to be laid on a bed of fill that will 'even out' any variations in the ground below.

GUIDANCE

Sometimes, fill capable of supporting a slab may already be present, albeit that it may well need to be rolled or compacted to improve its load bearing characteristics. However, in most cases it will be necessary for the ground to be overlain with 'hardcore', a controlled and selected type of fill specifically intended for placing around foundations or beneath ground bearing concrete slabs.

Materials for hardcore should be:—

- Chemically inert and free from any biodegradable material.

- Unaffected by water (i.e. no soluble or expansive materials).

- Sulfate free (sulfates — present in material such as gypsum plaster — can attack concrete and mortar).

- Free from toxic materials.

- Easy to consolidate and free draining, and with no pieces bigger than will pass through a 75mm diameter ring.

Well graded gravel, crushed rock, quarry waste, brick and concrete rubble, and blastfurnace slag have all been shown to be suitable.

After stripping topsoil and other organic material, at least 100mm of hardcore should be laid and compacted prior to finishing with a layer of 'blinding' (e.g. sand or weak concrete laid to a relatively smooth, level surface). Hardcore may be up to 600mm deep, providing it is compacted in 225mm thick layers. Attention should be paid to the backfilling of old excavations since excessive depth or a lack of compaction can lead to localised settlement. Where the depth of hardcore or fill is likely to exceed 600mm, a suspended floor structure should be used.

Ground bearing slabs must not be built-in to surrounding walls since differential settlement between the heavily loaded foundations and the relatively lightly loaded floor may cause the edges of the floor to 'shear' and crack. Redundant structures within the ground should be removed or the slab locally reinforced.

Poorly graded and excessively deep fill

COMMON MISTAKES

- Demolition rubble used for hardcore without sorting or grading (i.e. timber, plaster, etc. mixed with hardcore).

- Pieces of hardcore too large.

- Hardcore not specified to be formally compacted.

- Omission of blinding.

- Depth of hardcore greater than 600mm.

- Old structures in ground not removed or bridged.

REFERENCES

BS 8103–1:1995 cl.8

BRE Digests 276 & 427–1

BRE Report 332 pp.134–144

NHBC ch. 5.1–D4, S3, S4 & app. B

Ground Floors not Protected from Damp

Ground floors are particularly prone to damage as a result of exposure to moisture, especially those in direct contact with the inherently damp soil, or containing timber or timber products. The risk is increased if the site has a high water table (i.e. within 250mm of ground level), where surface water could penetrate the building (e.g. at the foot of a slope), or where the floor is close to ground level (e.g. where thresholds are designed for disabled access). Ground floors must be designed to prevent ground or surface water reaching the inside of the building or materials that are adversely affected by moisture.

GUIDANCE

Concrete floors in direct contact with the ground must incorporate an effective damp proof membrane (DPM), either above or below the slab; the former is preferable since the DPM will also protect finishes against construction moisture.

Sheets of low density polythene (LDPE) of at least 1200 gauge thickness (1000 gauge if the product is backed by a BBA Certificate or manufactured to PIFA Standards) is suitable as a membrane below the floor. The sheets should be laid on a smooth blinding (to prevent the membrane being punctured by hardcore) or on insulation boards. Joints should be lapped by at least 150mm and sealed with double-sided tape. The DPM should lap any adjacent damp proof course (DPC) by at least 75mm. If the membrane extends continuously into the surrounding walls (e.g. where it also serves as a radon barrier), it must be of a standard able to resist elongation and tearing (due to the settlement of the slab).

Polythene and bitumen sheet can also be used above the slab; also liquid materials such as 3 coat bitumen (2 coat where an air-dried thickness of 0.6mm can be achieved) or asphalt. Membranes on top of the slab must be protected by a screed or floating floor deck, other than where mastic asphalt (at least 16mm thick) is used.

No DPM is necessary if the floor is raised ('suspended') clear of the ground and away from the source of moisture, provided:—

• The ground (solum) beneath the floor is higher than that of the surrounding ground or it is effectively drained.

• Concrete floors are at least 100mm thick and contain a minimum of 300kg/m³ cement.

• The ground beneath a timber floor is covered to prevent plant growth (e.g. a 100mm concrete slab on hardcore or 50mm of weak blinding on a sheet membrane), and the void ventilated.

All types of suspended floor must be isolated from any part of the surrounding construction which could transmit moisture from the ground (e.g. the below ground portion of the inner leaf of a cavity wall or the sleeper walls supporting a timber floor).

Where high water tables or surface water present problems, subsoil drainage or some form of tanking may be necessary.

Perimeter of dpm lapping with dpc

COMMON MISTAKES

• Joints in sheet membranes not sealed or lapped.

• Membrane laid directly on hardcore (i.e. no blinding).

• Self adhesive bitumen sheet laid directly on polystyrene insulation boards (bitumen attacks polystyrene).

• No account taken of high water table or vulnerability of floor to surface water.

REFERENCES

BS 8102:1990 cl.4.1
BS CP 102:1973 cl.11
BRE DAS 35 & 36
BRE Digests 54 & 364
BRE GBG 28–1
BRE Report 332
pp. 83, 139–142 & 152–153
NHBC ch.5.1–D12, M8
NHBC ch.5.2–D6, D10 & D22
PIFA Standard 6/83A

Poorly Ventilated sub-Floor Voids

THE PROBLEM   RATING 3

Raising ('suspending') the structure of a floor is an effective way of isolating it from damp ground below. However, while the floor may not be in direct contact with the ground, it will still be in contact with the air within the void below, air which may itself be heavily laden with moisture, the result of the evaporation of water in the ground ('solum') beneath. For a timber floor, this can be as damaging as direct contact with the soil, and it is therefore essential that damp air is not allowed to build-up in the void below; hence the need for ventilation to be provided. It is also desirable that voids beneath concrete floors are ventilated, since any moisture infiltrating the concrete matrix will significantly increase the risk of reinforcement corroding (some manufacturers will not offer guarantees if the void is unvented).

GUIDANCE

At least two opposite sides of any void beneath a timber suspended floor must be ventilated. Ventilation should also be provided through any wall which sub-divides the void, including sleeper walls (care must be taken to ensure that the acoustic performance of any separating wall is not compromised). The amount of ventilation provided should be equivalent to at least 1500mm² per metre run of wall. This is usually achieved by an arrangement of airbricks, sometimes linked to a proprietary 'periscope' ventilator (e.g. where floor voids are below external ground level) or a run of ducting, such as where a ground bearing floor obstructs the perimeter of the void.

Voids below concrete suspended floors should also be vented on two opposite sides, with at least 600mm² of ventilation per metre run of wall (1500mm² if there is a gas service within the void).

There should always be some ventilation within 450mm of the corners of a void (to avoid stagnant pockets of air), and a gap of at least 150mm should be maintained between the underside of the joists (timber or concrete) or planks and the solum.

The number and spacing of airbricks will depend on their size, and the pattern and shape of their perforations. Details should be checked with manufacturers, although – for guidance – the following table sets out the standards of certain common sizes of airbrick manufactured to British Standards:–

Co-ordinating size of airbrick (mm)	Unobstructed air space (mm²)
225 x 75	1290
225 x 150	2580
225 x 225	4500

Airbricks for use in external walls (Class 1 units) may be made of concrete, clay, metal or plastic.

The need to ventilate a sub-floor void may also be dictated by the need to disperse radon, methane or carbon dioxide.

Typical 225x75mm clay airbrick

COMMON MISTAKES

- Ventilation paths blocked by separating walls or ground supported partitions.

- No ventilation where gas service runs in void below concrete floor.

- Airbricks specified in terms of size, with no account taken of actual amount of unobstructed air space.

REFERENCES

BS 493:1995 cl.7.4

BRE DAS 73

BRE Digest 364

BRE Report 332 pp. 81–82 & 91

NHBC ch.5.2–D10(b) & D22(c)

Inadequate Precautions against Gaseous Contaminants

THE PROBLEM   RATING **4**

Radon, methane and carbon dioxide are ground-based gases which must be treated as potentially harmful contaminants, since their presence within a building can have serious effects on the health and safety of the occupants: radioactive radon atoms attach themselves to airborne dust particles which, when inhaled, can induce lung cancer; certain concentrations of methane — which is flammable — can give rise to explosions; and carbon dioxide is a potentially lethal asphyxiant. However, this is not to say that the presence of these gases is always a problem; they only become dangerous when they exist above certain concentrations: 200 becquarels per cubic metre (Bq/m^3) for radon; 1% by volume in air for methane; and 5% volume in air for carbon dioxide (although precautions should actually be considered at 1.5%). Designers therefore need to be aware of the situations where gaseous contaminants may present a problem and what measures should be taken to prevent their internal build-up (entrapment).

GUIDANCE

Although the measurement of radon pre-construction is feasible, the results obtained are — on the basis of current knowledge — of little use in predicting the likely levels of radon that may become entrapped within a building. The only reliable method of assessing radon levels is to have long term measurements performed after construction. Reference must therefore be made to survey maps prepared by the National Radiological Protection Board (NRPB) or derived from their data.

Methane and carbon dioxide both result from organic decay, their presence being associated with — amongst other things — landfill sites (especially within 250m), rotting vegetation (including peat), coal measures at or near the surface and flooded mine workings. Various instruments and techniques can be used to measure their presence and concentration.

Having established that gaseous contaminants are present at levels which may be prejudicial to health, there are two basic ways of preventing their entrapment within a building:—

- **Primary protection:** Sealing the whole area of the ground floor with a gas proof membrane (which may need to be of a higher quality than a DPM) extending through all walls, across cavities and under partitions.

- **Secondary protection:** Ventilating below the ground floor (i.e. eliminating the pressure gradient between the soil and the building) by way of airbricks (suspended floors only), sumps connected to ventilation pipes rising to eaves level (which may be fan assisted), or by laying (solid) floors on open-textured fill that extends beyond the external walls.

Note that the BRE distinguishes between areas which require primary and secondary protection, those where secondary alone will suffice, and those where a risk assessment is required.

Where methane and carbon dioxide are present, monitoring should be continued over a period of time, usually two years.

Monitoring ground-based gases

COMMON MISTAKES

- Unsealed penetration of membranes by services.

- Membranes not taken across cavities in separating or external walls.

- Inadequate tear resistance of membranes where floors likely to settle differentially to walls.

REFERENCES

BS DD 175:1988

BRE Digest 395

BRE GBG 25

BRE IP 02/87

BRE Reports 211, 212, 332 & 376

CIRIA Reports 130, 131 & 149

NRPB Reports 254, 272 & 290

Inadequate Precautions against non-Gaseous Contaminants

THE PROBLEM
RATING **3**

Non-gaseous contaminants (e.g. heavy metals and toxic chemicals) are often present in the ground. Some are the result of a previous use (e.g. industrial processes), others occur naturally (arsenic or sulfates are widely found). Either way, it is important that they do not pose a threat to the health of occupants or cause the premature deterioration of building fabric. Ground treatment ('remediation'), or the protection of materials in contact with the ground may be required.

GUIDANCE

Contamination should be assessed against the "trigger/threshold values" set out in ICRCL Note 59/83. This will determine the necessity for remediation which can involve:

- **Removal:** Contaminated material is removed to a licensed tip by a licensed contractor (landfill tax may be payable).

- **Capping layers:** The contaminated material is overlaid with a barrier which reduces the possibility of it being reached. A 1000mm layer of clay covered with topsoil, or a thinner layer of clay on layers of hardcore/stone (both laid on a geotextile membrane) is generally sufficient for domestic gardens and play areas. Sometimes used after the relocation of contaminated material.

- **Bio remediation:** Micro organisms capable of ingesting the contamination are introduced, although only if there is sufficient time to allow the completion of the process (which must be monitored).

- **Soil washing:** The use of plant and machinery to rid the ground of the contaminants which are then removed from site. Used when removal would be prohibitively expensive.

- **Slurry trenches and membranes:** Vertical barriers to prevent contaminants migrating within, beyond or into the site (may also be required to prevent migration as a result of the use of a capping layer).

- **Vacuum extraction:** Pumping air into bore holes which is then extracted along with the contaminants.

- **Vitrification:** A current passed between buried electrodes melts and solidifies inorganic materials, which can then be removed. The more volatile contaminants are driven or burnt off.

Amongst the materials that may need protecting are concrete (acids and soft ground water can lead to sulfate attack) and plastics (chemical reactions can cause swelling, loss of strength and failure). The type and quality of cement in concrete should be related to the sulfate content of the ground, and special qualities of brick, block and mortar specified.

Spoil from an abandoned coal mine

COMMON MISTAKES

- No formal assessment of levels of contamination.

- Remediation disturbed by subsequent piling.

- Capping layers or barriers damaged by excavation for foundations or services.

- Contaminated soil spread indiscriminately across site.

- Cement content of concrete too low or inappropriate use of ordinary portland cement.

- Weak mortars specified for masonry below DPC level.

REFERENCES

BS 8301:1985 cl.5 & tab. 3
BS DD 175:1988
BRE Digests 363 & 395
BRE IP 02/87
CIRIA SP124

External Walls

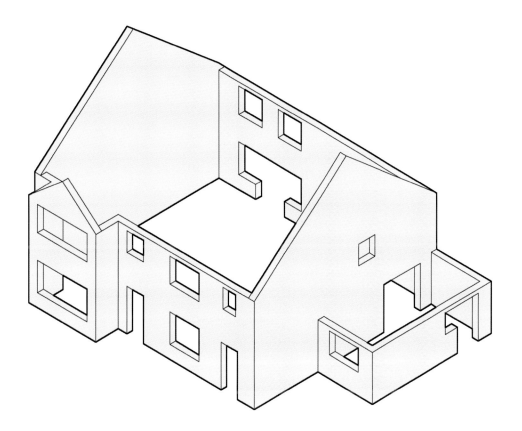

Defects and their Consequences

DEFECT	A	B	C	D
FAILURE TO CONSIDER STRENGTH AND STABILITY OF MASONRY WALLS	1	2		
INSUFFICIENT UNDERSTANDING OF TIMBER FRAMED STRUCTURES	3	1		
MASONRY WALLING NOT SUITED TO EXPOSURE		1	3	3
LACK OF PROVISION FOR MOVEMENT IN MASONRY WALLING	3	4		
MORTAR NOT MATCHED TO MASONRY OR CONDITIONS OF USE		3	2	3
WRONG TYPE OR TOO FEW WALL TIES	3	2		
INAPPROPRIATE USE OR POOR SPECIFICATION OF CAVITY INSULATION			3	
MISSING OR BADLY DETAILED DPCS AND CAVITY TRAYS			4	2
ILL-CONSIDERED DESIGN OF MASONRY PARAPETS	3	3	4	4
POORLY WEATHERED OPENINGS			3	4
AD HOC FIXING OF DOOR AND WINDOW FRAMES			2	3
DEFECTIVE INSTALLATION OF INSULATING GLASS UNITS				4
BADLY FIXED TIMBER OR PLASTIC CLADDING	2	3	1	2
INADEQUATE SPECIFICATION OF EXTERNAL RENDER		4	2	3

The four columns labelled A to D refer to the four types of consequence detailed on the opposite page.

KEY TO HAPM RATING

1 Low probability of defect occurring, and only likely to have minor consequences.

2 Low probability of defect occurring, though with potentially serious consequences.
   or
   Reasonable probability of defect occurring, though only likely to have minor consequences.

3 Reasonable probability of defect occurring and with potentially serious consequences.
   or
   High probability of defect occurring, though only likely to have minor consequences.

4 High probability of defect occurring and with potentially serious consequences.

Consequences

A

DISTORTION OF WALLING, COMPONENTS OR CLADDINGS

Walls are a key part of a building and any visible movement in them should be considered potentially serious. Bulging, leaning or buckling can be as indicative of defects in the design of a wall as it can of problems with foundations, roofs or floors. The 'binding' of doors and windows, the 'bowing' or twisting of components and — in extreme cases — cracked glass may all point to structural failure. Similarly the disruption of claddings and other components fixed to the walls.

B

CRACKING AND SPLITTING

Two main types of cracking occur in external walls: cracking due to structural movement (which runs with the horizontal and vertical planes of the mortar joints) and cracking due to changes in temperature or moisture content. The former types of crack often vary in width and run at oblique angles; cracks of the latter type are usually of uniform width and cut straight through materials at the weakest or least restrained part of a wall (e.g. above or below windows). Thermal movement can also split claddings and components. Other types of cracking may result from chemical action, sulfate attack being a common example (cracking also follows mortar joints, most noticeably in rendered walls).

C

INTERNAL STAINING, MOULD GROWTH AND FUNGAL DECAY

Damp penetration will generally result in the staining, discoloration and eventual breakdown of internal surfaces often — but not always — in the vicinity of the defect itself. On other occasions, the problem will be more widespread, with larger areas of damp that suggest a basic flaw in the design of the wall, the absence of an effective DPC perhaps, or an inappropriate use of materials. Condensation may also cause problems, including mould growth and the corrosion of ferrous materials (e.g. metal lath and plaster angle beads). High levels of moisture can ultimately lead to fungal decay (identification can be aided by the BRE's *Recognising wood rot and insect damage in buildings*), especially in skirtings or other timber in direct contact with the wall.

D

PREMATURE FAILURE OF COMPONENTS OR MATERIALS

Water, chemical agents (e.g. salts) and extremes of temperature or humidity can be a severe test of the materials and components from which an external wall is made, soon making evident shortcomings in its design. Spalled or friable masonry, decaying timber or metalwork, and the breakdown of plastics or sealants are all examples of damage that can be avoided by careful design and correct specification.

Failure to Consider Strength and Stability of Masonry Walls

THE PROBLEM   RATING **1 – 2**

The principles governing the structural design of simple, low-rise buildings have evolved over years of tradition and experience. Wall thickness and strength can be determined by 'rule of thumb', and without reference to a structural engineer. However, this approach is only valid within certain parameters, outside of which the structure may fail. It is therefore important that designers appreciate the limitations of this empirical approach to structural design, and when and where it may be used.

GUIDANCE

For houses or flats not exceeding three storeys (uniform floor loading 1.5kN/m², max. design wind speed 44m/s) and with no part higher than 15m above the lowest adjacent ground level (any difference in ground level on opposite sides of a wall must not exceed 4 times its thickness), the masonry of the external walls should have the following minimum strengths:—

- **One or two storey:** 5N/mm² brick or 2.8N/mm² block, except 7N/mm² (brick or block) below ground floor level if — on a two storey building — the height between top of foundation and the underside of the ground floor exceeds 1 metre.

- **Three storey:** 15N/mm² brick or 7N/mm² block up to first floor, 5N/mm² brick or 2.8N/mm² block above, except brick outer leaves to cavity walls to be 7N/mm².

Walls must not exceed 12m in height (measured to the vertical mid point of any gable) or be longer than 9m between vertical supports such as bonded returns or piers. The top of the ground floor must be no more than 1m above the top of the foundations, (2.7m where the floor provides lateral restraint). Roofs should span less than 12m, floors less than 6m; the area enclosed by the walls must not exceed 70m² if bounded on four sides, 36m² if bounded on three.

Leaves to cavity walls should be at least 90mm thick, solid walls at least 190mm thick (250mm if random rubble). Ground storey walls that do not receive lateral support from the ground floor and top storey walls with tall gables (i.e. more than about 3m) not tied at ceiling level must be proportionally thicker.

The combined length of openings in a wall must not exceed two-thirds of its length, with no narrow piers in between. No opening should be closer to the centre line of a vertical support than one-sixth of its width (an extra 150mm should be added at returns) and piers should be at least one-sixth of the total width of the openings either side; otherwise the structure must be proved by calculation. Wider piers may be necessary if the wall is loaded and masonry weaker than 7N/mm² is used. No opening should be wider than 3m. Lintels should have a minimum bearing of 150mm in the plane of the wall or 100mm at right angles, unless supporting a concrete floor in which case the bearing should be one-tenth of the span or 150mm, whichever is the greater.

Corbelling out from a cavity wall

COMMON MISTAKES

- Walls outside of simple rules not formally designed by a structural engineer.

- Supporting partitions not tied or bonded to walls.

- No account taken of ground floors that are significantly higher than ground level.

- Stability of narrow piers not justified by calculation.

- Introduction of 'solid wall' features such as corbels and arches into cavity walls, but without reinforcement or additional support.

REFERENCES

BS 5628–3:1985 cl.27.6 & 27.8

BS 8103–2:1996 cls.4–6

BRE Report 352 pp.16–23

Insufficient Understanding of Timber Framed Structures

THE PROBLEM

RATING **1 – 3**

Although the structural function of a timber framed wall is the same as that of a masonry wall, the way in which it works is quite different. The relatively slender structural members (studs) and the need for wall panels to be able to resist forces acting within their plane ('racking') means that subjects such as the stability at openings, the provision of sheathings and linings, and the frequency and size of fixings are of prime importance. Designers must have a clear understanding of what is involved in a timber framed structure if they are to make informed decisions at the outset, even though they may not be responsible for its detailed design (which should be by a structural engineer or a specialist fabricator).

GUIDANCE

Most new timber framed housing in Britain and Ireland consists of storey-height wall panels with each floor providing an erection platform for the next. Other methods of construction include balloon framing (panels rise to roof level in one) and 'post and beam' (as seen on historic buildings). Factory made box units forming complete rooms for craning into place are also used.

First storey of a timber framed house

Wall panels are framed from timber studs, typically at 400mm or 600mm centres, and twice-nailed to a head and cill rail. Studs are sized as compression members liable to bending in accordance with BS 5268–2:1991, taking into account the interaction between the studs and any sheathing. Timber lintels must be provided above openings, other than those which are not subject to any direct load; lintels are fixed between studs with additional 'cripple' studs beneath their ends. Noggings to support fixtures and fittings should also be provided.

A sole plate — fixed down and protected by a DPC — should be used to transfer loads to the foundations and to provide a level, accurate base on which to erect the ground floor panels. The sole plate may need to be secured against overturning or sliding in areas where wind loads are high.

Racking resistance is usually achieved by sheathing the frames with a sheet material such as 12.5mm bitumen impregnated insulation board or 9.5mm plywood, fixed with 3mm diameter x 50mm long nails spaced at 300 or 150mm centres, depending on the type of material (perimeter fixings should be at half these spacings). An external masonry veneer may — if suitably tied to the timber frame — contribute up to 25% of the required racking resistance, a plasterboard lining (fixed with 2.65mm diameter x 40mm long nails at 150mm centres) up to 50%.

Floors and roofs must be able to transfer lateral loads (i.e. act as diaphragms), generally meaning that flooring must be directly fixed to joists, and that roofs should be fully braced and have a plasterboard ceiling. Joists and rafters must bear directly on studs (or be offset by no more than their width) unless additional head binders — justified by calculation — are provided along the tops of the wall panels.

COMMON MISTAKES

- Absence of cripple studs.
- Cripple studs and ordinary studs not securely connected to ensure composite action.
- Sole plate not fixed down or levelled to provide a square, accurate base.
- No fixing schedules provided for use on site.
- Reliance on masonry veneers or plasterboard to provide 100% of racking resistance.

REFERENCES

BS 5268–6.1:1996
BRE Report 233 pp.4, 19 & 22
TRADA WIS 0–03
TRADA 1994 pp.27–38

Masonry Walling not Suited to Exposure

THE PROBLEM   RATING 1 – 3

The ability of a masonry wall to resist rain penetration is dependent on the thickness and quality of its materials, and the existence of any cavity or external finish (i.e. render or cladding), all of which must be suited to the 'exposure' of the wall to wind-driven rain. Failure to take account of this fact can result in not only water penetration but also the accelerated decay of the masonry; walls must be designed with due regard to their exposure if their long-term performance and endurance is to be ensured.

GUIDANCE

The exposure of the wall should be assessed in accordance with the *local spell index* method in BS 8104:1992 (formerly DD93) or by reference to the simplified exposure zones defined in BRE Report 262. Higher categories should be assumed if there is any doubt.

Cavity walls with a minimum external leaf thickness of 90mm and a 50mm clear cavity are suitable for most situations, although it may be prudent to increase both at higher exposures. Wider cavities should also be used if both leaves are of block (mortar droppings cannot be easily cleared if the cavity is any narrower).

Solid walls should be:–

* Rendered or clad if the exposure is greater than 56 litres/m^2 (moderate/severe), and always clad at very severe exposures.

* If rendered: at least 90mm in sheltered locations, 190mm in sheltered/moderate and 215mm in moderate/severe; severe exposures require 328mm clay or calcium silicate, or 250mm dense concrete (lightweight concrete can be thinner).

* If unrendered: 328mm thick if sheltered, otherwise 440mm.

External leaves and solid walls should be constructed of:–

* Type FL, FN, ML or MN clay bricks to BS 3921:1985. Type F if there is any risk of saturation (e.g. below DPC and for cills) and type L (low salt content) below DPC or where the wall is rendered and sulfate resisting cement is not used.

* Class 2 to 7 calcium silicate bricks to BS 187:1987. No class 2 must be used below DPC and only class 4 or above for cills.

* Concrete blocks or bricks to BS 6073–1:1981. Blocks below DPC should be dense ($\geq$1500kg/m^3), made with dense aggregate or have a strength of at least 7N/mm^2. Bricks should be at least 7N/mm^2 (15N/mm^2 if likely to be saturated); 20N/mm^2 bricks should be used below DPC and 30N/mm^2 for cills etc.

Stone can be selected on the basis of BRE Reports 29, 36, 84 & 134, local experience and the advice of individual quarries. Cills, copings, etc. of limestone or sandstone must to be 'edge' bedded, meaning that stones longer than the depth of the natural bed cannot be specified. Cast stone should be to BS 1217:1997, reconstructed stone to BS 6457:1984.

Vulnerable recessed jointing

COMMON MISTAKES

* Design of walls not based on assessment of exposure.

* Narrow cavities to walls that have two blockwork leaves.

* Use of 'O' quality clay bricks for rendering.

* Recessed jointing between units (will encourage water penetration).

REFERENCES

BS 5390:1976 cl.20

BS 5628–3:1985 cls.21.2–3 & 22

BS 8104:1992

BRE Digests 420 & 441

BRE Report 262 p.22

BDA DN 16 pp.4–13

HAPM TN 02

Lack of Provision for Movement in Masonry Walling

THE PROBLEM   RATING **3 – 4**

All types of masonry expand or contract in response to variations in their temperature or moist-
ure content. Sometimes this movement is reversible. At other times it is not, especially that which
occurs during the period immediately following construction when the wall 'dries out', causing
the shrinkage of concrete or calcium silicate masonry and the expansion of clay brickwork, and
– potentially – the formation of cracks, especially where the wall is unrestrained or lightly
loaded. Hence the need for masonry walling to be designed to accommodate horizontal movement.

GUIDANCE

Long runs of masonry (i.e. where not restrained by a return, a bonded loadbearing partition or a cross wall), should be broken into panels by vertical joints at the following *maximum* centres:–

- **Clay brickwork:** 12 metres (assuming an average expansion of 1mm per metre of wall), though up to 15 metres for storey-height walls that are fully loaded or restrained.

- **Calcium silicate brickwork:** 7.5 to 9 metres subject to no panel being longer than three times its height.

- **Concrete block or brickwork:** 6 metres subject to no panel being longer than twice its height; applies also to cast stone *and* to the inner leaves of cavity walls, especially if built of autoclaved aerated blocks (slip ties should be provided across the joint) or the leaf is not subject to much load.

- **Natural stonework:** 12 to 15 metres for ashlar walling (the fine joints – rarely wider than 4mm – mean that spalled edges and displacement can occur) if the movement cannot be accommodated within the mortar, otherwise unnecessary.

The distance between an external or internal angle and the first joint should be no more than half the general spacing, preferably less if the angle is fully bonded. Similarly, joints in parapets or other walls that are completely unrestrained. A joint should also be provided at the angle if the return is less than 800mm long (675mm if the walls are formally designed by an engineer).

Under certain circumstances, it may be possible to space joints at greater centres than those stated above or even to omit them altogether, taking advantage of bricks or blocks that can be shown to undergo only minor amounts of movement and the flexibility of weak, lime-based mortars. Specific product manufacturers should be consulted and the likely movement checked by calculation.

Movement joints should also be considered where there is any change in thickness or height (e.g where a single and two storey wall meet), and where areas of masonry above or below openings that have a significantly smaller vertical cross sect-ion than that of the wall either side (these panels will tend to crack unless a movement joint or bed reinforcement is provided).

Movement joint at return in a brick wall

COMMON MISTAKES

- Movement joints considered in plan, but not in elevation (i.e. no account taken of lengths of walls above or below openings).

- Extra joints not provided to parapets and similar lightly loaded panels.

- No movement joints at short returns, or at changes in height, loading, etc.

- Continuity of joints not maintained at cills, lintels, decorative projections, etc.

REFERENCES

BS 5390:1976 cls.21.1 & 21.2

BS 5628-3:1985 cl.20.3

BS 8103-2:1996 App. B

BRE DAS 18

BRE Digests 157, 361 & 441

BRE Report 352 pp.24–27

BDA DN 10

Mortar not Matched to Masonry or Conditions of Use

THE PROBLEM

RATING **2 – 3**

Mortar is used in masonry walling for many reasons. It binds together the bricks, blocks or stones, prevents wind and rain penetration and helps distribute loads; it must be strong (but not too strong), durable and easy to use (workable), qualities that can only be achieved by selecting an appropriate combination of binders (cement or lime), aggregates (usually sand) and additives. An understanding of mortar and care in its specification are essential if the performance of the wall is to be ensured.

GUIDANCE

Mortar mixes (in terms of relative volumes of dry materials):–

Class	Cement:lime:sand	Masonry cement:sand	Cement:sand + plasticizer
(i)	1:0 to 1/4:3	n/a	n/a
(ii)	1:½:4 to 4½	1:2½ to 3½	1:3 to 4
(iii)	1:1:5 to 6	1:4 to 5	1:5 to 6
(iv)	1:2:8 to 9	1:5½ to 6½	1:7 to 8

Mixes within the same class have the same strength, though cement:lime mortars have improved adhesion (and hence greater resistance to rain penetration) than air entrained mixes. Class (i) mixes – which are relatively stiff and impervious – have the best strength and durability though little ability to accommodate movement. Finely graded sands are used at the lower proportions, coarse sands at the upper; the latter are essential when laying engineering bricks, or where water penetration must be resisted.

Plasticizers introduce (entrain) air into a mix, increasing its workability and resistance to freezing. They are also used with masonry cement (either pre-mixed or introduced on site) and may be added to cement:lime mortars. However, they must be used with care, since too much air will reduce strength and durability (the air content of the mortar should be between 10% and 18%).

Recommended mixes (always weaker than the masonry):–

* Clay bricks can be laid in class (i), (ii) or (iii); class (iii) must not be used below DPC or with 'N' quality clay bricks intended for rendering; class (i) should be used for cills, copings, etc.

* Calcium silicate bricks and concrete blocks can be laid in class (iii) or (iv), with class (ii) or (iii) below DPC and class (ii) for cills, copings, etc; class (iv) should not be used with concrete bricks or where the walling is at risk from saturation.

* Stonework is laid in mixes weaker than class (iv), sometimes without cement and using aggregates other than sand; some sandstones and granites require stiffer, stronger mortars; class (iv) mixes can be used for cills, copings, etc.

A 'general purpose' air entrained 1:1:5½ cement:lime mortar mix can also be used (but not with stone) provided low permeability and exceptionally high or low strengths are not required.

Typical air-entraining plasticizer

COMMON MISTAKES

* Specification does not make clear that materials are to be batched dry.

* Grading of sand not fully considered or specified.

* Sulfate resisting cement not used where 'N' quality clay bricks are used below DPC or intended for rendering.

* No control in the use of air entraining admixtures.

* Strong mixes not used for cills, copings, etc.

* Excessively strong mortar used for stonework.

REFERENCES

BS 5628–3:1985 cl.23
BRE DAS 70, 71 & 72
BRE Digest 362
BRE IP 10/93
BRE Report 352 pp.81–82

Wrong Type or Too Few Wall Ties

THE PROBLEM   RATING **2 – 3**

The structural stability of a cavity wall depends on its leaves being securely tied together. Each leaf stiffens the other, preventing buckling and distributing lateral loads, such as those generated by the push and pull of the wind. A timber or metal frame stiffens a masonry facing in a similar manner (though not vice versa — the frame is designed to stand by itself). Wall ties capable of transmitting tensile and compressive forces must therefore be used in sufficient numbers if the failure of the outer leaf or — at worst — the collapse of the wall are to be avoided.

GUIDANCE

Wall ties can be classified by shape (to BS 1243:1978 – there are three patterns; butterfly, double triangle or vertical twist) or stiffness (to DD 140-1:1987 — types 1 to 4 for masonry walls, and types 5 and 6 for masonry claddings to timber frame construction).

For masonry walls, select ties to suit the width of the cavity:—

Cavity (mm)	Shape to BS 1243:1978	Type to DD 140-2:1987
75 or less	Butterfly, double triangle or vertical twist	1, 2, 3 or 4
76 to 90	Double triangle or vertical twist	1 or 2
91 to 100	Double triangle* or vertical twist	1 or 2
101 or over	Vertical twist	1 or 2

tested to satisfy DD 140 type 2

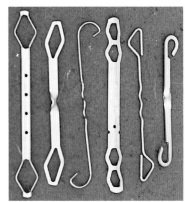

A selection of DD140 masonry wall ties

COMMON MISTAKES

- Lightweight ties used for cavities wider than 90mm.
- Embedment less than 50mm.
- No account taken of extra width of cavity when partial fill insulation used.
- Ties in timber framed walls fixed to sheathing not studs.
- Additional ties not provided along sloping verges, either side of movement joints or at openings.

Ties should be long enough to ensure a minimum of 50mm embedment in each leaf and that drips are set centrally within the cavity; longer ties may be needed if the cavity is 'offset' by the use of partial fill insulation. They should be set in a staggered pattern and spaced no further apart than 450mm vertically and 900mm horizontally (600mm for certain types of partial fill insulation board; check manufacturer's details). Ties to timber framed walls should be fixed to the studs at 375mm or 525mm vertical centres (600mm or 400mm stud spacing respectively).

Walls should be provided with extra ties at 300mm vertical centres and within 225mm of:—

- The vertical edges of openings, and at unreturned ends.
- Sloping edges to verges.
- Each side of vertical movement joints.

Ties either side of movement joints should be able to cater for some horizontal force, and ties should not be placed too close to returns. Some manufacturers of autoclaved aerated concrete blocks caution against the use of rigid ties due to the risk of the high initial shrinkage of the blockwork being transferred to the outer leaf.

REFERENCES

BS 5268–6.1 cl.4.10
BS 5628–1:1992 cls.29.1.4–6
BS 5628–3:1985 cl.19.5
BS 8103–2:1996 cl.6.6
BRE DAS 115
BRE IP 4/81, 6/86, 16/88 & 17/88
BRE Report 233 pp.21 & 30
BDA DN 15 p.13
TRADA 1994 p.93

Inappropriate Use or Poor Specification of Cavity Insulation

Demands for energy conservation require external walls to have levels of thermal performance that are not readily achievable by masonry alone. Insulation has to be added, often between the leaves of a cavity wall. However, since the cavity exists to protect the inner leaf of the wall from moisture, filling it with insulation can defeat its purpose. Either the cavity must be maintained, or use made of insulation systems that do not readily transmit water. Damp penetration can only be avoided if designers appreciate the dangers of filling a cavity and take great care in their specification and detailing.

GUIDANCE

Suitability of cavity insulation: Firstly, the exposure of the wall should be assessed in accordance with the *local spell index* method in BS 8104:1992 (formerly DD93) or by reference to the simplified exposure zones defined in BRE Report 262. Higher categories should be assumed if there is any doubt. Then it should be established if the wall is to be clad or rendered and — if not — what type of jointing (i.e. flush or recessed) will be used and how cills, copings, etc. will be detailed. Finally, the proposed insulation and its method of installation (built-in or injected) is assessed in the context of its third party certification (all cavity insulation systems must be backed by a certificate issued by the British Board of Agrément or other European Technical Approval issuing body).

Badly fitted partial fill cavity insulation

COMMON MISTAKES

- Selection of insulation not based on formal assessment of exposure.
- Wall ties not spaced to suit partial fill insulation.
- Manufacturing tolerances of boards or batts not taken into account when deciding width of cavity.
- Inspection procedures for blown or injected insulation not formally specified.
- Full fill cavity batts used as partial fill insulation (danger of slumping and bridging of the cavity).

The likely standard of workmanship that can be expected on site should also be taken into account. Cavity insulation should not be specified if high standards cannot be achieved.

In general terms, the use of cladding or render will allow cavity insulation to be used over a wide range of conditions, whereas features such as recessed joints or flush cills will limit its use to sheltered — sometimes very sheltered — exposures.

Partial fill insulation: A 50mm wide open cavity should be maintained between the insulation and the outer leaf, and wall ties spaced so as to fully support every board (typically 600mm horizontal centres supporting 1200mm wide panels). Ties must be of 'eccentric' drip design, with purpose-designed retaining clips.

Full fill insulation: Presents the greatest risk of rain penetration and hence needs the highest standard of workmanship. Batts must be held in place by the masonry but never compressed; ideally, the nominal width of the cavity should be slightly increased and its permitted tolerances made explicit (the manufacturing tolerance of a 100mm thick mineral wool batt is −0mm/+15mm). Blown and injected systems require formal inspection and approval procedures which must be clearly specified at the outset.

Detailing: Cavity trays *may* be required above insulation that is stopped short (e.g. at the ceiling in a gable), though beware of creating slip planes in walls exposed to high winds.

REFERENCES

BS 5628–3:1985 cl.21.3.2.8
BS 6676–2:1986 cls.3 & 5
BRE GBG 35
BRE IP 12/94
BRE Report 262 pp.21–25
BDA GPN CIW–2

Missing or Badly Detailed DPCs and Cavity Trays

THE PROBLEM   RATING 2 – 4

Cavity walls cannot absorb and hold moisture in the same way as thick, heavy walls made of soft, porous materials. Interiors can only be kept dry if they are designed as barriers to moisture, whether falling from above, driven sideways or rising from the ground. Damp-proof courses (DPCs) and cavity trays are therefore critical components – second only in importance to the walling itself – and designers must be clear as to their function, where they are necessary and how they should be detailed.

GUIDANCE

The need for DPCs in a cavity wall is based on the assumption that water will penetrate the outer leaf and run down its inner face, as well as rising upwards from the ground due to capillary action.

Each leaf of a cavity wall must have a DPC immediately above ground level. The outer DPC must be at least 150mm above finished external levels, though the inner – which must overlap with any floor DPM by at least 75mm – can be lower, provided there is a 225mm clear cavity below (150mm if the foundation is a raft and the base of the cavity is drained); if this is not possible (e.g. when there is a significant difference between floor and ground levels), vertical DPCs and perhaps tanking will have to be considered.

Inner and outer leaf DPCs do *not* need to be connected unless the cavity needs to be sealed against gaseous contamination, in which case it must be ensured that cavity trays do not slope inwards.

DPCs should also be provided:–

• Under pervious or jointed cills.

• Vertically at jambs so as to separate the outer and inner parts of the wall. DPCs should extend at least 25mm into the adjacent cavity and lap over any DPC below the cill. Certain types of proprietary cavity closer may be used in lieu.

• Over openings, unless well protected by an overhang or similar. Cavity trays (lapped over jamb DPCs) should be used.

Trays should also be provided above any place where the cavity is bridged and at abutments, ensuring that dpcs, flashings and roof coverings are considered together as a system.

Cavity trays should be drained by weepholes at 900mm maximum centres (closer if the cavity is to be filled and saturation of the insulant avoided), with at least two weepholes per tray. Trays should be fully supported and rise from outside to inside by at least 150mm and – if not continuous – extend 25mm beyond any vertical DPC and have properly formed 'stopped' ends to prevent water draining into the cavity. Pre-formed trays are preferred.

Joints in flexible DPCs and trays should be lapped by at least 100mm. Laps in DPCs that must resist downward-moving water or hydrostatic pressure *must* be sealed.

Omission of cavity tray over opening

COMMON MISTAKES

• Unnecessary use of cavity trays to link DPCs (makes installation more difficult than it need be).

• Inward sloping cavity trays to link DPCs at two levels.

• Vertical DPCs too narrow.

• No cavity trays over small openings such as ducts and recessed meter boxes.

• Too few weepholes above cavity trays.

• Cavity trays unsupported.

• No stopends to cavity trays above openings etc.

• Laps in DPCs and cavity trays not specified to be sealed.

REFERENCES

BS 5628–3:1985 cls.21.4 & 21.5

BS 8215:1991

BRE DAS 35, 36 & 98

BRE Digest 380

BRE GBG 33

BDA DN 16 pp.14–22

NHBC ch.5.1 D–12

Ill-Considered Design of Masonry Parapets

Parapets are one of the most vulnerable elements of an external wall. Being high up and lacking the protection of a roof, they are exposed on two sides and from above, which leaves them particularly susceptible to the adverse effects of the weather. Structural instability, the accelerated deterioration of materials and water penetration are all issues that have to be addressed in the design of masonry parapets, with particular care being taken to ensure that copings and cappings — 'the first line of defence' — are weather resistant and not readily displaced.

GUIDANCE

Stability: For buildings not exceeding three storeys in height and where access to roofs is limited, masonry parapet walls should comply with the following table (height is from where the wall meets the underside of the structural roof to the top of the capping or coping; for cavity walls, thickness is the combined thickness of both leaves; no parapet must be wider than the wall below):–

Height	Solid wall thickness (mm)	Cavity wall thickness (mm)
600mm or less	150 min.	200 min.
600–760mm	190 min.	225 min. (interpolated)
760–860mm	215 min.	250 min.

Masonry: Parapets should be built in type FL or FN clay bricks (ML or MN if rendered), class 3 to 7 calcium silicate bricks, concrete bricks with a strength of at least 20N/mm², dense concrete blocks (i.e. at least 1500kg/m³ or made with dense aggregate), concrete blocks of at least 7N/mm² strength or certain types of aerated concrete block. Solid parapets must *not* be rendered both sides.

Mortar: Clay brickwork should be laid in a class (i) or (ii) mortar, though class (iii) can be used if rendered and type 'L' bricks are used; sulfate resisting cement is necessary for type 'N' bricks. All other types of masonry can be laid in a class (iii) mix, other than unrendered concrete blockwork, which should be laid in class (ii).

Cappings and copings: Preformed copings (ideally sloping inwards) or those built up on creasing tiles should have an overhang of at least 40mm; cappings ('flush' copings) are not advised in other than moderate or sheltered exposures. Class (i) mortars should be used for bedding clay units, class (ii) otherwise. Thought should be given to using L-shaped or 'clip over' copings and to the use of dowels or joggles where there is any risk of displacement.

DPCs and cavity trays: A continuously supported DPC (with 100mm sealed laps) is required beneath all jointed copings or cappings (in the case of the latter, positioned two courses down in order to add weight to the DPC). In a cavity wall, a drained cavity tray linked with the roof flashing to form a continuous barrier to water should be provided at the base of the parapet; stepped trays to sloping parapets need stop ends.

Displacement of coping to raked parapet

COMMON MISTAKES

- Sulfate resisting cement not used where parapet built of 'N' quality bricks.

- Flush copings used in severe or very severe exposures.

- Copings or cappings not fixed in place where at risk from displacement.

- No support to DPC below coping or capping.

- Cavity trays not stepped by at least 150mm.

- Omission of stop ends to stepped cavity trays.

REFERENCES

BS 5628–3:1985 cls.21.5, 21.7 & 22

BS 8103–2:1996 cl.9

BS 8215:1991 cl.5.5

BRE DAS 106

BRE GBG 33

BDA DN 16 pp.21–21

Poorly Weathered Openings

THE PROBLEM   RATING 3 – 4

The installation of doors and windows in a masonry or masonry-clad wall involves the disruption of cavities and the formation of sometimes difficult junctions, presenting designers with a series of problems regarding their positioning, sealing and protection. Water penetration and the premature deterioration of materials or components can be a serious issue if any of these factors are not adequately addressed. Designers should always try to produce solutions that have been proven by experience, which preferably do not place too much reliance on the use of sealants.

GUIDANCE

Door and window frames should be set back from the face of the wall by at least 50mm. Ideally, they should be within the 'dry' part of the wall (i.e. across the cavity or within the inner leaf), which lends protection and makes it easier to fully point the reveals.

Cills should be bedded on mortar and extend to project 40–50mm beyond the face of the wall, though they should not be so wide that they 'cup' or distort; a secondary 'sub' cill or threshold is preferable to an excessively wide primary cill; ideally, a primary cill without a sub-cill should have horns built into the surrounding masonry.

Sub-cills (preferably of unjointed precast concrete or stone to BS 5642–1:1978) must always be provided where the exposure is severe, with the door and window frames set back (rebated) behind the masonry outer leaf. A water bar should connect the two cills.

The method of weatherproofing the junction between the wall and the door or window frames depends on whether they are built into or fitted to prepared openings at a later date.

If built-in, the jamb DPC is attached to the frame and folded back ready for the raising of the wall. The gap forward of the DPC is filled with sealant or a pre-compressed impregnated foam tape (the latter is preferable since it allows trapped water to evaporate). For wooden frames, this method is most suited to windows that are set back and not in contact with damp masonry.

Where windows are installed into preformed openings, a gap of between 5 and 15mm must be left between the jambs and the masonry. This prevents the frame touching the 'wet' part of the wall and permits a sealant or foam joint to be formed with ease (the DPC must still project enough to ensure the total separation of the outer and inner leaves and to master the lugs of metal or plastic frames).

Sealants should have a min. face width of 10mm (triangular) or 6mm (rectangular) and a depth no less than their width.

Level access thresholds are exceptionally vulnerable to penetration by driven water (threshold seals are unreliable and should not generally be used). Such doorways should be protected by a porch or lobby in all but sheltered locations.

Over-reliance on sealant pointing

COMMON MISTAKES

- Timber cills too wide and hence prone to 'cupping'.
- No built-in horns where cills or thresholds used without sub-cills.
- Sub-cills or rebated reveals not provided in severe or very severe environments.
- Timber windows in direct contact with damp masonry.
- Gaps around frames greater than 15mm (too wide to fill under most circumstances).

REFERENCES

BS 5628–3:1985 cls.21.5.4, 22.1.1
BRE DAS 68
BRE Report 352 p.172
BPF 348/2 pp.16 & 17
BPF 356/1 pp.12–22 & 28
HAPM TN09
SWA FS 03
TRADA 1993 pp.35–39

Ad Hoc Fixing of Door and Window Frames

THE PROBLEM   RATING **2 – 3**

Door and window frames — timber, plastic or metal — must be prevented from moving in any way that could cause their displacement or distortion (e.g. twisting) if doors and opening lights are to function with ease and remain weatherproof. This means that they must be fixed to walls in a way that resists the lateral force of the wind and the dynamic loads associated with opening and closing. Failure to clearly specify how frames must be fixed may lead to uncertainty and potentially defective installation.

GUIDANCE

Frames that are built-in to masonry should be secured by angle cramps fixed to their backs and built into the mortar joints as the walls are raised; screwing the frames into plugs set in the wall is an acceptable alternative. Doors and windows in timber framed walls are pre-fixed to the frame and not to any masonry cladding.

Door frames and un-glazed windows can also be screw-fixed to the reveals of pre-formed openings (behind either the glazing line or an opening light), although this may cause difficulties where the jamb fixings coincide with an insulated cavity closer. Brackets that fix back to the inner leaf can be used to overcome the problem; lugged brackets are required for PVC-U or metal windows, and can also be used with specially profiled timber frames (brackets are nearly always used when installing frames with pre-glazed fixed lights). Timber frames must have a 10mm clearance all round to allow for positioning and adjusting fixings. The profiles of PVC-U and metal frames permit the use of smaller gaps, though note that PVC-U windows require an edge clearance of at least 5mm to accommodate thermal expansion (8mm if of a dark colour).

Frames should be fixed as follows (min. two fixings per jamb):

- **Timber:** Within 150mm of cills, thresholds and heads, and at 450mm centres along jambs; heads to frames over 1500mm wide should also be fixed.

- **PVC-U:** Within 250mm of corners and junctions (but no less than 150mm), and at 600mm centres to jambs, heads and cills; head fixings may be further apart if profiles are reinforced.

- **Steel:** Within 190mm of corners and at approximately 650mm centres along jambs. Heads and cills are fixed where frames are wider than 1000mm. Thresholds and heads to doors should also be fixed within 190mm of corners.

- **Aluminium:** Within 150–200mm of corners and at 500mm centres to jambs, heads and cills.

To prevent damage due to over tightening and to maintain clearances, packing pieces should be provided at all fixings. Cills must not be punctured by fixings (brackets must be used) and special thought needs to be given where fixing to lintels; special clips or foam-fixing might be required.

PVC-U frame fixed to inner leaf

COMMON MISTAKES

- Frames nailed to masonry, not screwed or cramped.
- Gap between frame and wall too small to permit fixing.
- Fixings too far apart.
- PVC-U frames fixed close to corners (i.e. within 150mm).
- Need for packing pieces not made explicit.
- Cills 'fixed through' to wall.

REFERENCES

BS 8000–5:1990 cl.3.5.6
BPF 348/2 pp.11–13
BPF 356/1 pp.23–27
SWA FS 03
TRADA 1993 pp.36 & 37

Defective Installation of Insulating Glass Units

THE PROBLEM   RATING 4

Condensation ('misting') between the panes of insulating glazing units in timber frames is a common problem, often within a very short time of installation. It is generally a clear indication that the seals around the edge of the unit have broken down due to saturation (sealants swell and detach from the glass when exposed to excessive water), exposure to ultraviolet light or contact with aggressive chemicals. Units are also heavy and if not properly installed will move as doors and windows are opened, creating stresses that can damage opening lights, often causing them to let in water. Many variables are involved, making a well thought-out specification essential if problems are to be avoided.

GUIDANCE

Insulating glass units (IGUs) must comply with BS 5713:1979 (this number must be marked on the spacer bars, otherwise it is *not* a BS unit), preferably ones that are kitemarked too. Dual-seal units are superior to single-seal, with silicone based secondary seals being preferable to polysulphide or polyurethane. Edge tapes can trap water and — if present — must be removed before installation.

Spacer bars contain a desiccant which absorbs any moisture trapped within the unit. However, the desiccant can only absorb about 20% of its dry weight and hence enough must be available if it is to remain effective. 12mm bars with frequent, adequately sized holes contain about the right amount for an average domestic window (i.e. 20mm thick units if 4mm glass is used), though 20mm bars contain even more. 6mm bars are only suitable for very small units and should be avoided.

Setting and location blocks hold the IGU firmly in place and maintain clearances (3mm all round for units up to 2m long, though 6mm at the bottom if drained and ventilated). Blocks must be at least 25mm long x the thickness of the unit, wedge-shaped if used on sloping rebate platforms and provided in accordance with section 4.2 of the GGF Manual.

Distance pieces must be provided between the face of the glass and the bead or rebate upstand (50mm from corners and at 300mm centres) unless load-bearing glazing tapes are used.

Glazing systems can be fully bedded or drained and ventilated. Full bedding involves using a single, low permeability sealant (typically low-modulus silicone), a vapour permeable sealant externally with a low permeability one inside, or various types of load-bearing tape. Drained systems — the best option — rely on air flowing around the edge of the IGU and drainage slots in the bottom bead.

Rebates and glazing beads should be 18mm high if units are fully bedded, 22mm if drained and ventilated; beads should be wider than they are high, and the bottom bead should extend beyond the rebate and be weathered to shed water. The glazing platform should be deep enough to accommodate the IGU, beads and 3mm clearance front and back (5mm for larger panes).

Condensation between the panes of an IGU

COMMON MISTAKES

- Use of IGUs not manufactured to British Standards.
- Edge tapes not removed prior to installation.
- Spacer bars too narrow for the size of unit.
- No setting or location blocks and distance pieces.
- Units bedded in linseed oil putty or butyl-based 'non setting' compound.
- Chemical incompatibility of sealants (with each other or with decorative coatings).
- Rebates or beads too small for thickness of unit.

REFERENCES

BS 6262:1982 cls.5.3–2, 7.1–4 & 9
BRE Report 28
GGF 4.2, 4.2.1 & 5.16
HAPM TN 12

Badly Fixed Timber or Plastic Cladding

THE PROBLEM   RATING **1 – 3**

How a cladding is fixed is as important an aspect of its performance as the selection of the cladding material itself. Wind-induced suction may lift a cladding from its supports and dimensional changes – the result of variations in temperature or moisture content – can open up joints to water penetration, exposing the boards and the structure behind to excessive wetting. Support battens, the frequency and type of fixing, and the accommodation of movement all have to be considered if the integrity – and hence durability – of the cladding is to be maintained.

GUIDANCE

Timber boards and plastic planks (cellular PVC to BS 7619:1993) are fixed to 19 x 38mm treated softwood battens spaced at max. 600 centres, creating a drained and ventilated gap between the cladding and the wall behind; this is to limit condensation, permit water to escape and – in the case of timber cladding – prevent cupping by allowing the boards to dry equally via both faces.

Battens are fixed at 90° to the line of the cladding profile, with additional battens at corners and jambs; plastic claddings may also require battens to support horizontal edges. Horizontal battens should be weathered so as to direct moisture outward, with gaps left between their ends (unless 'board on board' cladding or counter battens are used). Vertical battens to timber frames must coincide with studs; horizontal battens that span counter battens or studs without sheathing must be thicker than 19mm (generally twice the board thickness to ensure a suitable depth for nailing).

Horizontal timber boards: Single ('secret') fixing for tongued and grooved boards, plus shiplap or rebated boards less than 100mm wide; two fixings per board otherwise. Tongues should engage by at least 12mm and laps be no less than 15mm. An 8–10mm gap must be left at the end of each board to allow for drying out (assuming a moisture content of 18% ± 2% at fixing) and future maintenance. Joints must always occur at a batten, a double batten if large amounts of horizontal movement are anticipated.

Vertical timber boards: Double fixing for boards wider than 75mm. Outer boards to 'board on board' cladding must be fixed heart side out and nailed through the gaps between the inner boards, which are fixed heart side in.

Plastic planks: Each locked to the previous plank and nailed to the battens through their tongued edge, working from one end or out from the centre to avoid 'springing' between fixing points; vertical starter planks are nailed through both edges. Edge and angle trims are fixed at 600mm centres, starter trims (horizontal) and drip trims (vertical) at 300mm. Expansion gaps should be to manufacturer's recommendations.

Stainless steel lost-head, annular ring shank nails are preferred for fixing, except for cedar boards (round head nails should be used).

Profiled plastic cladding to a dormer

COMMON MISTAKES

- No gap behind cladding (e.g. boards or planks fixed direct to studwork).
- Battens too thin for nailing.
- Tongued and grooved timber boards wider than 100mm.
- Board-on-board claddings fixed with heart side facing the wrong way.
- Timber boards nailed one through the other.
- Plastic planks 'sprung' into place or between fixings.
- No provision for thermal or moisture movements.

REFERENCES

BPF 349/1
TRADA WIS 1-20
TRADA 1994 pp.97–102

Inadequate Specification of External Render

THE PROBLEM   RATING 2 – 4

External renders can significantly enhance the weather resistance of a wall, as shown by their extensive use over many years in areas with a high degree of exposure to wind-driven rain. However, their performance is highly dependent on the correct preparation of the surface to be rendered, the number and thickness of coats used, and the selection of a mix that is durable yet not prone to cracking. Any rendering specification must cover these points if the rapid deterioration of the render is to be avoided.

GUIDANCE

Preparation: 10–12mm recessed joints will provide a sufficient key to brickwork, though engineering brick — like all smooth, dense surfaces — should be coated with a strong cement:sand slurry ('spatterdash' or 'stipple') or overlaid with metal lath; the latter is usually necessary for stonework and always required over sheets, boards or insulation. Textured surfaces (e.g. hammered concrete or most concrete blocks) may not need preparing.

Number of coats: For masonry or concrete with an exposure of less than 100 litres/m² (determined using the local spell index method set out in BS 8104:1992 or by reference to BRE Report 262), a two coat render specification is acceptable; a three coat specification is used for higher exposures or on metal lath.

Thickness: Undercoats on masonry of concrete should be 6–12mm thick with final coats specified to be just thicker than the maximum size of sand particle, typically 6–7mm but sometimes as little as 3mm or as great as 10mm. In three coat work, the first undercoat should be slightly thicker than the second. On metal lath, the first undercoat should be 3–6mm thick, the second 10–14mm. Two coat renders should have a total thickness of no less than 16mm, three coat ones no less than 20mm.

Mix: The following table shows a range of mixes (batched by volume on the basis of *damp* sand) designated in BS 5262:–

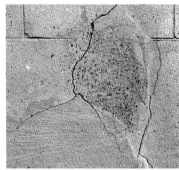

A botched repair of a cement-rich render

COMMON MISTAKES

- Undercoat too thick (local irregularities in the backing should be 'dubbed out' and allowed to dry before the application of the render).
- Specification not clear that mixes are based on volume of damp sand (if dry sand is used, cement content should be increased by 15 to 20%).
- Type and grading of sand not specified.
- Sulfate resisting cement not specified where soluble salts might be present.
- Final coat too cement-rich or sand too fine (both will result in surface cracking).

Type	Cement:lime:sand	Cement:sand + plasticizer	Masonry cement:sand
I	1:¼:3	n/a	n/a
II	1:½:4 to 4½	1:3 to 4	1:2½ to 3½
III	1:1:5 to 6	1:5 to 6	1:4 to 5
IV	1:2:8 to 9	1:7 to 8	1:5½ to 6½

Type II mixes are suitable for undercoats where the final coat is to be thrown (i.e. wet or dry dash), III if trowelled or floated, though on low density concrete (including aircrete blocks), type III and IV mixes should be used. A type I first coat should be used on metal lath. Type II, III or IV mixes can be used for final coats, though only type II in areas of high exposure (unless the use of a type III undercoat is dictated by the backing). Sand should be clean and sharp with a good mix of coarse and fine particles; sulfate resisting cement should be used if there is any risk of salts being present.

REFERENCES

BS 5262:1991 cls.17–19 & 31–34
BS 8104:1992
BRE Digest 410
BRE GBG 18
BRE Report 262 p.22

Roofs

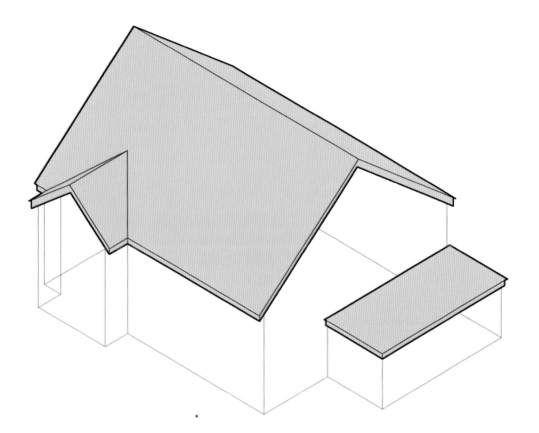

Defects and their Consequences

DEFECT	CONSEQUENCES			
	A	B	C	D
INAPPROPRIATE OR ILL CONSIDERED USE OF TRUSSED RAFTERS			2	2
TRUSSED RAFTER ROOFS INADEQUATELY TIED OR BRACED			2	1
LACK OF CONNECTION BETWEEN TIMBER ROOFS AND WALLS			2	1
FLAT ROOFS OR LINED GUTTERS NOT DESIGNED TO FALLS	3			
INSUFFICIENT VENTILATION OF COLD ROOFS	3			
LAP OF SLATES OR TILES NOT SUITED TO PITCH	3			
EXCESSIVE LENGTHS OF SHEET METAL COVERINGS OR LININGS	2	3		
SHEET METAL LAID ON UNSUITABLE SUBSTRATE	2	2		
INADEQUATE SPECIFICATION OF ASPHALT OR BUILT-UP FELT ROOFS	4	3		
ROOF COVERING NOT SECURED AGAINST WIND UPLIFT	2	2	3	
POOR DETAILING OF PITCHED VALLEYS TO SLATED AND TILED ROOFS	4		2	
INADEQUATE SPECIFICATION OF METAL FLASHINGS	3			

The four columns labelled A to D refer to the four types of consequence detailed on the opposite page.

KEY TO HAPM RATING

1 Low probability of defect occurring, and only likely to have minor consequences.

2 Low probability of defect occurring, though with potentially serious consequences.
or
Reasonable probability of defect occurring, though only likely to have minor consequences.

3 Reasonable probability of defect occurring and with potentially serious consequences.
or
High probability of defect occurring, though only likely to have minor consequences.

4 High probability of defect occurring and with potentially serious consequences.

Consequences

A   WATER STAINING, MOULD GROWTH AND FUNGAL DECAY

Protecting the interior of a building from the weather is the fundamental purpose of any roof. Water penetration due to defects in roof design can stain, discolour and cause the deterioration of internal finishes at high and low level (penetrating water will always try to run downwards). Poor design can also encourage condensation within the roof, leading to the growth of mould. What is more, high levels of moisture will invariably result in fungal decay and insect attack (identification can be aided by the BRE's *Recognising wood rot and insect damage in buildings*), an extremely serious problem in an element where timber is often a key structural component.

B   DETERIORATION OF COVERINGS

Roof coverings are subject to extremes of temperature and wetting, and if incorrectly designed may prematurely deteriorate or break down. Spalled, cracked or missing slates and tiles, split or corroded sheet metal (note that 'underside' corrosion may not even be visible to start with), and blistered or ruckled bituminous coverings are all indicative of the early failure of poorly designed roofing systems, even if water penetration had yet to occur.

C   DEFORMATION OR DISPLACEMENT OF ROOF

Sagging ridges or pitches, hogging over separating walls or other primary structural elements and the 'uplift' of verges all suggest defects in the design of a roof. Likewise dislodged slates or tiles, distorted or detached flashings, and the opening-up of joints in sheet metal roofs. Although failures of this sort are rarely 'catastrophic' (the possible exception being wind damage), the often progressive nature of their occurrence almost always results in a need for major repairs.

D   CRACKING AND DISTORTION OF WALLS

Defects in the design of the roof may result in the structural movement of external and internal walls. Leaning, bulging, cracks, the rotation of wall plates and the disruption of cover flashings may all be symptoms of such problems which — in extreme cases — can lead to complete detachment and collapse (e.g. of gable walls).

Inappropriate or Ill-Considered Use of Trussed Rafters

THE PROBLEM   RATING **2**

The use of trussed rafters is an efficient and economic way of framing pitched roofs. Lightweight trusses comprising stress-graded timber members joined by punched metal plates are made-up in a factory to suit the span and profile of the roof, the fabricator taking responsibility for the design of timbers and connectors. Used at close centres, they combine the functions of common rafter and ceiling joist in one, eliminating the need for separate purlins, binders or ridge boards. However, trussed rafter construction does have limitations and building designers must be aware that it is they — not the truss fabricator — who are responsible for many basic decisions including the profile, span, pitch and spacing of the trusses, and the size and position of water tanks, hatches, rooflights etc. Failure to consider these factors at the outset can seriously compromise the practicality of trussed rafters, resulting in ad-hoc modifications on site and the weakening of the whole roof structure.

GUIDANCE

Trussed rafters are best suited to roofs that are simple in form and plan, where the pitch is the same throughout and not interrupted by rooflights, chimneys or other openings. Anything more complex will invariably necessitate the design of a variety of types and size of truss, the introduction of cut timbers to 'make-up' parts of the roof and special bracing. Ideally, decisions on the use of trussed rafters should be guided by the following points:—

Typical trussed rafter roof structure

COMMON MISTAKES

- Spans should not exceed 12 metres.

- Trusses should be spaced evenly and no further apart than 600mm between centres. Additional trusses will be necessary if the standard spacing does not result in a truss within 50mm of gable or separating walls (both sides).

- Hatches, rooflights, etc. should be accommodated within the rafter spacing. If this is not possible, the distance between the trimming trusses may be increased to twice their design spacing. Otherwise groups of compound (multiple) trusses supporting cut purlins and ceiling binders will be required.

- Supporting walls should always coincide with a truss connector plate which, in turn, must overlap the wall plate by at least 50% and not oversail it by more than 50mm.

- The method of formation of hips and valleys is carefully considered in relation to their weatherproofing function.

- The practicality of supporting the necessary size of water tanks (for duopitch trusses, tanks should generally be supported centrally on bearers; small span trusses may require the removal of up to two trusses and the insertion of a portion of cut roof).

- Design responsibilities not clearly understood or agreed at the outset.

- Roofs too complex for effective use of trussed rafters.

- No truss adjacent to gable or separating wall.

- No extra trusses to trim wide hatches, rooflights, etc.

- Hips or valleys not carefully thought-out.

- No consideration of position or loading of water tanks.

Since trussed rafters cannot be readily altered, their use might not be appropriate if there is any possibility of the roof being converted to a room sometime in the future (unless use can be made of purpose-designed 'attic' trusses).

REFERENCES

BS 5268–3:1998 cls.6.2, 7 & 11
BS 8103–3:1993, cl.7
TRA SFG G12 pp.17–21
TRADA WIS 1–10 & 1–29

Trussed Rafter Roofs Inadequately Tied or Braced

The integrity of a trussed rafter roof is dependent on the rafters being held rigidly in position. Any buckling (due to vertical loading) or racking (due to wind loading) will give rise to lateral forces for which neither the roof nor its supporting walls are likely to have been designed, potentially leading to the deformation and collapse of the whole structure. Bracing is therefore essential and building designers must appreciate when and where different types of bracing are required (note that it is generally the building designer that is responsible for the stability of the roof, not the truss supplier).

GUIDANCE

For low-rise buildings in areas other than open or level country with no shelter (e.g. fens, moorland, etc.) with roof spans up to 12m, a floor to ceiling height no greater than 2.6m and a maximum unsupported wall length of 9m, the following bracing specification may be adopted:–

- **Longitudinal** bracing (i.e. at right angles to the rafters) at all unsupported joints (nodes), and finishing tight against a separating wall or gable. Bracing at rafter level is fixed to the webs of the trusses (permits fixing of diagonal bracing).

- **Diagonal** bracing fixed to the underside of the rafters at an angle of approximately 45°. The braces run from ridge to wall plate and repeat over the length of the roof (min. four braces to be used). May be omitted from up to two trusses between braces and one truss at separating or gable walls.

- **Chevron** bracing is fixed to certain webs where duopitch roofs span more than 8m and monopitches more than 5. Extra webs are braced where spans exceed 11m and 8m respectively; also the faces of monopitch roofs that are not restrained by a wall or cladding. Each brace must connect at least three trusses.

Bracing should be at least 89mm wide x 22mm deep (min. cross-sectional area 2134mm²) and twice-nailed to every truss. Joints in bracing should be lapped over at least two trusses. Tiling battens (max. 400mm centres x min. 1200mm long) must be fixed *directly* to the rafters (i.e. 'warm' roofs may require extra bracing) with no more than 1 in 4 joined over a single truss. Diagonal, chevron and longitudinal braces at rafter level can be omitted if rigid sarking is used. Plasterboard ceilings should also be provided throughout, otherwise diagonal bracing to the bottom chords.

Specially designed bracing will be required if the trusses are more than 600mm apart, the ceiling follows the pitch of the rafter ('room in the roof'), truss nodes are not in line or chord depths vary, cantilevered or stub trusses are proposed, or the bracing is required to stabilise the walls ('wind girders'). Additional bracing may also be required at hips and valleys due to absence of tiling battens. Ties may also be required where areas of the roof are 'made-up' with cut rafters.

Longitudinal and diagonal bracing

COMMON MISTAKES

- Positions of access hatches, rooflights etc. do not take account of bracing.

- Longitudinal braces stopped short of masonry walls.

- No additional bracing where insulation is fixed between rafters and tiling battens.

- No account taken of rafter level bracing when planning a 'room in the roof'.

- No additional bracing to bottom chords in absence of plasterboard ceiling.

- Unstable areas of cut roof (e.g. at hips or valleys).

REFERENCES

BS 5268–3:1998 cl.7.2, App.A
BS 8103–3:1996 cl.7.4 & Annex H
BRE GBG 8
BRE Report 302, pp.66 & 67
TRA SFG G12 pp.8–16
TRADA WIS 1–10 & 1–29

Lack of Connection between Timber Roofs and Walls

All parts of a roof structure (including gables) are vulnerable to damage by the effects of the wind, suction forces being especially a problem. These can be large enough to lift a roof off its supports or to cause masonry gables to topple outwards, particularly in exposed locations (masonry has little inherent ability to resist the tensile forces that arise from the lateral pull of the wind). It is therefore essential that roof members are securely connected to wall plates — which may need to be connected to the walls below — and that masonry gables are firmly tied (strapped) to a well-braced roof structure.

GUIDANCE

Every rafter (cut or trussed) or joist in a roof must be positively connected to the bearing surface, typically a 70–75mm timber wall plate. Framing anchors, clips or nails (min. 2no. 4.5mm x 100mm long skew-nailed from each side) can be used, although nailing is not advisable for trussed rafters due to the risk of damage to the connector plates. The bearing must be long enough to accommodate the required fixings (this is especially critical for flat roof joists where — from a structural point of view — bearings as short as 24mm may be acceptable).

The roof must be connected to the supporting walls below with vertical straps at centres not exceeding 2m unless:—

- It has a pitch of 15° or more, a covering that weighs at least 50kg/m² (e.g. concrete or clay tiles), is of a type known by local experience to be resistant to wind gusts, and where the main members (i.e. the rafters of a single roof, the purlins or trusses of a double roof) are no more than 1.2m apart.
- It is part of a timber-framed building.
- It can be demonstrated by calculation that the downward forces due to the dead load of the roof are at least 1.4 times the likely maximum uplift force.

Straps must be no less than 1m in length and 2.5mm thick, and have cross-sectional area of at least 75mm². Either they should be fixed to the face of the masonry or have 100mm tails built into a bed joint. Screws are preferable, although nails may be used (but not into brickwork or mortar joints, or for flat roofs). Each strap should be fixed at least 4 times, with two fixings engaging the last full block and one within 150mm of the end of the strap. The tops of the straps should be turned over the top of the wall plate and thrice-nailed, or twisted and fixed to the sides of the rafters.

Gable walls must be strapped to roofs at 2m centres (max.) along the pitch, each strap (min. cross-sectional area of 150mm²) being long enough to be fixed four times to the third rafter from the wall (noggings are fitted between the rafters). Ends of straps are built-in and turned-down the face of the masonry by 100mm. Straps may be required at ceiling level if the roof pitch is about 20° or more.

Fixing of trussed rafter to wall plate

COMMON MISTAKES

- Rafters not clipped or nailed to wall plates.
- Length of bearing too short to accommodate fixings.
- Vertical straps too far apart or less than 1.0m in length.
- Straps inadequately fixed or fixings not clearly specified.
- No noggings to rafters where connected to gable wall.
- Gap between last rafter and masonry not packed where straps connect roof to wall.
- End of straps to gable wall turned-down over insulation rather than face of masonry.

REFERENCES

BS 5268-3:1998 cls.7.2.4 & 7.3
BS 8103-1:1995 cl.5.4
BS 8103-3:1996 cls.6.2, 6.4 & 7.10
BRE GBG 16
TRA SFG G12 pp.28–31

Flat Roofs or Lined Gutters not Designed to Falls

THE PROBLEM   RATING **3**

Water, being a liquid, cannot support itself as a body and hence should not — in theory — be able to remain on an unrestrained flat surface. However, this is not the case in reality where the truly 'flat' roof or lined gutter is neither practical nor functional. Friction, surface tension (a physical characteristic of all liquids) and deflection due to load can all result in rainwater 'ponding' on 'flat' surfaces, often exposing them to long periods of continuous wetting. This will rapidly expose any weaknesses in design or construction (e.g. capillary action at unsealed laps — as found in fully-supported metal sheet roofing). Hence the reason why flat roofs and lined gutters must always be designed to 'fall' towards an eaves or an outlet.

GUIDANCE

Flat roofs and lined gutters should be constructed to the following minimum finished falls at any point:—

Covering	Pitch	Angle (degrees)	Rise over 1.0m
Asphalt, Built-up bitumen and Lead	1:80	0.75	13.0mm
Copper and Zinc	1:60	1.00	17.5mm

However, to allow for construction inaccuracies and for deflections, the roofs or gutters should be *designed* to at least twice these falls (i.e. 1:40 or 1:30). The following should also be noted:—

- If falls in different directions intersect, the minimum fall should be maintained at the mitre (i.e. falls generally will be greater than the minimum).
- The falls should be such as to shed water in the direction of the gutters or rainwater outlets.
- Particular care is needed in detailing falls around rooflights and other obstructions.

Falls in the deck of a flat roof can be achieved in a number of ways, the most common being:—

- **For timber joisted structures:** tapered joists or tapered timber firrings; the latter should have a minimum thickness of 20mm, more if required to span across joists.
- **For concrete decks:** screeds laid to falls, generally with a minimum thickness of 25mm and a maximum of 200mm, although some proprietary lightweight screeds may be designed for use outside of these limits. Screeds should be laid in bays not exceeding 3000mm in any direction (to control shrinkage cracking).

Designers must ensure that they allow sufficient depth to accommodate the required falls and — in the case of sheet metal roof coverings — stepped drips (55–60mm for lead, 65mm for copper and zinc).

Ponding due to inadequate falls

COMMON MISTAKES

- Roofs designed to 'minimum' falls only.
- No account taken of cross falls or intersections.
- Drainage outlets not placed at lowest point of roof.
- Insufficient depth allowed for falls or drips.

REFERENCES

BS 6229:1982 cls.15.1 to 15.5
BS CP 143–5:1964, cl.304
BS CP 143–12:1970 cls.3.2.5–6
BRE Digest 419
BRE Report 302 p.58
CDA TN 32
LSA 1992 pp.26 & 104
ZDA 1988

Insufficient Ventilation of Cold Roofs

A 'cold' roof comes into being when insulation is placed below any part of its structure (pitched or flat), typically at ceiling level and with a void over (it should be noted that the roof is 'cold' even if the insulation is wholly within the thickness of the structure, such as when it is placed between rafters). The fact that the insulation exists to prevent heat loss means that when the external air temperature is low, so too is the air temperature above (or even within) the insulation. But low temperature air can very easily become saturated with moisture, at which point condensation can form on adjacent cold surfaces (the lower the temperature, the less moisture the air can hold and vice versa) and condensation within a roof can cause considerable damage, specially to timbers (risk of fungal decay) and ceilings. It is therefore essential that moisture rising up through a building is not allowed to build up on the 'cold' side of the insulation. Either it must be prevented from reaching the cold side of the insulation (rarely easy to achieve in practice) or it must be removed. Hence the need for cold roofs to be ventilated.

GUIDANCE

The void above the insulation in a cold roof should be ventilated at eaves level via openings to the external air on *opposite* sides of the roof (i.e. the roof must be *cross* ventilated). The minimum amount of ventilation provided should be *equivalent* to:–

- For roofs with a pitch of 15° or more, a 10mm clear strip running full length (i.e. 10,000mm² per metre of eaves).

- For simple rectangular roofs with a pitch of less than 15° and a span of less than 10 metres, or where the pitch is 15° or more and the ceiling – and thus the insulation – follows the pitch of the roof (e.g. where there is a 'room' in the roof), a 25mm clear strip running full length (i.e. 25,000mm² per metre of eaves).

- For complex roofs with a pitch of less than 15° or a span in excess of 10 metres, 0.6% of the plan area of the roof (i.e. 6000mm² per square metre).

Extra ventilation – equivalent to a 5mm clear strip – should be provided at ridge level where the roof has a pitch of 35° or more or its span is in excess of 10 metres, at the head of any monopitch roof, and to pitched roofs where the ceiling follows the slope. In the case of the latter – and flat roofs too – there should be at least 50mm of free air between the insulation and the deck or covering. High-level ventilation should never be used on its own (risk of water vapour being drawn *into* the roof).

Care should be taken to ensure that air paths are not in any way obstructed (e.g. by separating walls or cavity barriers; alternative provision may need to be made) and that the configuration of the roof does not inadvertently lead to stagnant pockets of air.

Note: Although reference is made to cold flat roofs, their use is strongly discouraged, especially if the roof abuts a wall or is surrounded by a parapet.

Roof slope vents behind a parapet

COMMON MISTAKES

- No free air space between insulation and roof covering.

- Proprietary roof slope vents set too far apart or too high.

- Ventilation paths obstructed by noggings, fire barriers, separating walls, etc.

- Configuration or roof results in stagnant pockets of air, such as where gables interrupt eaves.

- Over-fascia vents blocked where sarking dressed to gutter.

REFERENCES

BS 5250:1989 cls.3.8 & 9.4

BS 5534–1:1997 cl.25.5.1

BS 6229:1982

BRE Report 262 pp.3–7, 17 & 18

BRE Report 302 pp.27, 84 & 155

Lap of Slates or Tiles not Suited to Pitch

THE PROBLEM   RATING **3**

Slates and tiles are laid in lapped courses so that there are no direct paths through which water can pass. However, rainwater does not always fall vertically and run down the roof under the force of gravity. It is often driven over the plane of the roof by the wind, and up and under the overlapping units. The lap between successive courses of slates or tiles must therefore be large enough to prevent driven water from penetrating as far as their top edges ('heads') and spilling over into the building.

GUIDANCE

Slates and tiles are laid double or single lap. In the case of the former, the lap is the amount that the tail of a slate or tile covers the head of one in the *next course but one* below; there are *three* thicknesses of slate or tile at the lap. In single lap coverings the tails of one course lap the head of the course *immediately below*, leaving only *two* thicknesses at the lap.

Slates — natural and artificial — should be fixed with the following *minimum* headlaps (the actual lap depends on the size and quality of the slate; the advice of individual producers or manufacturers should always be sought):—

Single-lap profiled tiles with side locks

Rafter pitch (degrees)	Lap: moderate exposure (mm)	Lap: severe exposure (mm)	Gauge (distance between nails)
45 plus	65	65–70	For centre nailing equals $\dfrac{\text{length of slate} - \text{lap}}{2}$
40	65	75–90	
35	75	75–110	
30	75	75–115	For head nailing equals $\dfrac{\text{length of slate} - (\text{lap} + 25)}{2}$
25	90	100–120	
20	115–130	130+	

Only large (min. 500mm long x 250mm wide) centre-nailed slates should be used below 30° (although offering better protection to the holes, large slates should not be head-nailed). If slates are laid to diminishing courses, the lap must remain uniform; it is the length of the slate that varies.

Plain tiles should be laid double-lap at a pitch of at least 40° with a minimum lap of 65mm for moderate exposures, and 75mm for severe. The lap should never exceed a third of the length of the tile. Some types of concrete plain tile may be used at 35°.

Single-lap tiles (e.g. pantiles) are typically laid at pitches of between 30° and 47.5° with headlaps of 75 (min.) to 100mm and side laps of 50mm. The lap may also be dictated by grooves ('side locks') moulded to suit the manufacturer's recommended minimum pitch, which may be as low as 15°.

Note: It is never advisable to use any type of slate or tile at its recommended minimum pitch; add at least 5° to allow for inaccuracies in construction and extraordinary weather.

COMMON MISTAKES

- Insufficient lap for exposure and pitch of roof.
- Battens gauged to suit length of rafter, not the required lap.
- Small slates used at below 30° rafter pitch.
- Lap of head-nailed slates not measured from hole.
- Single lap tiles used at absolute minimum pitch.

REFERENCES

BS 5534–1:1997 cls.3.3 & 3.4
BRE DAS 142
BRE Report 302 pp.68 & 97

Excessive Lengths of Sheet Metal Coverings or Linings

THE PROBLEM   RATING **2 – 3**

The prolonged life and easy workability of lead, copper and — more recently — zinc make them eminently suitable for covering roofs and lining gutters. However, metals expand and contract in response to changes in temperature, the stresses induced often being large enough to split metal sheeting. Hence the reason why sheet metal coverings and linings must not be used in long lengths (stress is proportional to strain is proportional to length).

GUIDANCE

Lead sheet roofing and linings should be as follows:—

BS Code	Max. centres of joints with fall (m)	Max overall girth of gutters (m)	Max. between drips for gutters & roofs up to 10° pitch (m)	Max. between laps for roofs between 10°–80° pitch (m)
4	0.500	0.750	1.50	1.50
5	0.600	0.800	2.00	2.00
6	0.675	0.850	2.25	2.25
7	0.675	0.900	2.50	2.40 (2.25 at 60°+)
8	0.750	1.000	3.00	2.50 (2.25 at 60°+)

Rolled joints are used with the fall, although welts and standing seams can be used for steep pitches. Joints across the fall comprise either drips or laps (roofs over 10° only). Splash laps should be omitted from steps where the pitch is less than 3°.

Copper sheet roofing (1.8x0.60 or 0.75m sheets) and linings should be laid to the following maximum sizes. Standing seams — as opposed to rolls — are only used above a 6° pitch. Joints across the fall comprise either double welts (single used over 45°) or drips.

Laying a zinc strip roof

COMMON MISTAKES

- Locations of joints not decided at design stage.
- Joints spaced too far apart.
- Lapped or welted joints used at low pitches.
- Splash laps included to steps in gutter linings.

Thickness (mm)	Width between standing seams (m)	Width between rolls (m)	Unbroken length of gutter (m)
0.45 or 0.60	0.525	0.500	1.500 to 4.500
0.70	0.675	0.650	4.500

Roofs using hard tempered copper strip can be laid in bays of up to 8.5m (0.60m wide for normal conditions, 0.45m for exposed).

Zinc roofing and linings (strips or 2.438 metre long sheets) should be laid to the following lengths (50° maximum pitch). Sheet edges are jointed by either capped rolls or standing seams (over 5°). Steps and welts are used across the fall (welts only over 15°).

Unworked width of zinc strip or sheet (m)	0.65 mm	Thickness 0.70 mm	0.80 mm
0.500	n/a	10	12
0.610	6	9	10
0.686	6	9	9
0.914 (strip & sheet)	4	6 (sheet only)	6 (sheet only)
1.000	n/a	4	6

REFERENCES

BS 6915:1988 cls.5.1 & 5.3
BS CP 143–5:1964 cl.405
BS CP 143–12:1970 cl.3.4
CDA TN 32 p.65
LSA 1992 pp.26, 64 & 102
ZDA 1988

Sheet Metal Laid on Unsuitable Substrate

THE PROBLEM   RATING **2**

The successful performance of a fully supported sheet metal roof covering or gutter lining is as much dependent on the substrate on which it is laid as it is the quality and detailing of the metal sheet itself. A poorly designed substrate might prevent the covering from moving freely, leading to levels of thermal stress for which it was not designed. Or it may leave the vulnerable underside of the metal sheeting exposed to water or aggressive acids, risking the onset of underside corrosion (it is often forgotten that 'raw' metals — especially lead and zinc — are readily attacked by water; durability only arises after the formation of a protective coating due to weathering). Designers of fully supported sheet metal roofs and gutters must therefore allow for the interaction between the coverings or linings and its substrate if premature failure of the roof is to be avoided.

GUIDANCE

Sheet metal coverings or linings can be laid on timber or concrete substrates. They should not be laid direct on insulation due to the risk of water being drawn through the laps as a result of the rapid cooling of air trapped beneath the metal sheeting (thermal 'pumping'); where 'warm' construction is desirable (e.g. flat roofs), the sheet metal should be laid on a separate, batten-supported timber deck, the resultant space being ventilated to the external air. Substrates must be dry before laying is commenced.

Timber substrates — which should be at least 19mm thick — may comprise either boarding or sheet materials such as exterior grade plywood (e.g. WBP). Boards may be tongued and grooved or plain edged; the former are less likely to curl and cause 'ridging', although the latter — laid with gaps in between — may be preferable if there is a risk of condensation on the underside of the metal (the maximum width of the gaps is determined by the thickness of the metal sheeting — e.g. 5mm for codes 5 and 6 lead, 10mm for codes 7 and 8). The corners and external angles of all boards or sheets should be rounded (especially important with plain edged boarding) and all fixings punched or countersunk below the surface. Boarding should be laid diagonally or in the direction of the fall.

Concrete substrates should be smooth and dry; the surface should be sealed with a hard-drying bitumen coating if drying times are likely to be prolonged.

Underlays isolate the sheet metal from the substrate. Their purpose is to ensure the metal can freely expand and contract, and to mitigate any possibility of premature corrosion resulting from the direct contact of incompatible materials (e.g. copper and ferrous fixings or lead and concrete). Impregnated felt (Type 4A(ii) No. 2 inodorous to BS 747), building paper to BS 1521 Class A (smooth decks only) and polyester geotextile (non-woven needle-punched with a weight of not less than $210g/m^2 \pm 5\%$) are all suitable. Geotextile should always be used if there is a risk of condensation forming immediately below the metal.

Underside corrosion of lead sheet roofing

COMMON MISTAKES

- Sheet metal laid direct on insulation.

- No provision for ensuring substrate is dry before sheet metal is laid (includes drying of preservative treatments).

- Boards fixed at right angles to the fall of the roof.

- No underlay or wrong type of underlay used.

REFERENCES

BS 747:1994 cl.5

BS 6915:1988 cl.4

BS CP 143–5:1964 cls.204 & 404

BS CP 143–12:1970 cls.3.2.1–4

CDA TN 32 pp.8–9

LSA 1992 pp.9 & 12–15

LSA 1993

ZDA 1998

Inadequate Specification of Asphalt or Built-up Felt Roofs

Flat roofs covered in mastic asphalt or built-up bituminous felts can provide many years of trouble-free service. But they are extremely unforgiving of poor design and specification, the dark colour and low mass of the materials having only limited tolerance to variations in surface temperature. The nature of the substrate on which the roofing is laid, the need for separating layers, the formation of perimeter details and service penetrations, and the provision of solar protection must all be carefully considered if asphalt or bitumen sheet covered roofs are to endure without leaking.

GUIDANCE

Substrate (deck): Only warm-deck construction should be used. The 'sandwich' type (insulation below the waterproof membrane) is preferable for timber decks (min. 19mm thick), the 'inverted' type (insulation above the membrane) for concrete. Mastic asphalt — which is brittle at low temperatures — is best laid on concrete or other heavyweight decks (polymer-modified asphalt *must* be used if laid on timber or insulant). Concrete or screeded decks must be thoroughly dry before roofing; temporary drain holes or vents might be required. Independent kerbs must be provided where timber decks abut vertical or sloping surfaces.

Insulation boards: Must be rigid and thermally stable (i.e. low coefficient of thermal expansion or of composite construction) and — if used for inverted roofs — able to withstand extremes of temperature and wetting. On sandwich roofs, the boards are bonded to a vapour control layer, typically comprising one or two sheets of Type 5U (to BS 747) bituminous felt. The edges of the vapour control layer should be turned-back 150mm over the insulation and bonded to the roofing membrane.

Mastic asphalt: Type R988 (to BS 6925) laid on sheathing felt (Type 4A to BS 747) in two coats to a minimum thickness of 20mm. Angles must be filleted and upstands keyed (e.g. raked joints or metal lath).

Bituminous membranes: Built-up from layers of oxidised felts to BS 747: first (venting) layer — Type 3G loose-laid; second layer (underlay) — Type 5U bonded in hot bitumen; third (capping) layer — Type 5B or E also bonded in hot bitumen). Note that the venting layer is perforated and hence only partially bonded (reduces thermal stresses) and that Type 5 (polyester) felts are superior to Type 3 (fibreglass). BBA approved polymer-modified felt laid in two or three layers may also be used, and proprietary single-ply bituminous or polymeric membranes are also available; reference should be made to manufacturer's literature and the terms of any third party certification. Fillets (45°) must be provided at upstands.

Surface protection: Except for Type E or other mineral-faced felts, 6mm stone chippings or lightweight tiles bedded in bitumen compound, or solar reflective paint (less durable). Membranes on inverted roofs are protected by the insulation and slabs or ballast.

Built-up bituminous felt roof covering

COMMON MISTAKES

- No allowance for drying of concrete before laying roof.
- Coverings carried across dissimilar substrates.
- No vapour barrier between deck and insulation.
- Vapour barriers not turned up and bonded to roof finish.
- No separating layer beneath mastic asphalt.
- Upstands at abutments etc. dressed at acute angles.
- No solar protection to upstands.

REFERENCES

BS 747:1994 cls.4 & 6

BS 6925 1988 cl.3

BS 8217:1994 cls 5–8

BS 8218:1998 cls.6.2–6.7

BRE Digests 312, 324, 372. & 419

BRE GBG 36

BRE IP 8/91 & 7/95

BRE Report 302 pp.136–154

MAC 1999

Roof Covering not Secured against Wind Uplift

THE PROBLEM   RATING 2 – 3

When wind is deflected and accelerated over the roof of a building it produces in its lee a zone of suction which increases in strength as the pitch of the roof decreases. The forces induced are often great enough to dislodge slates, tiles and sheet metal coverings, with ridges, hips, eaves and verges being particularly susceptible to damage (due to the localised increase in suction where the wind whips up and over the edges of the roof). Designers must ensure that specifications for the fixing of roof coverings are devised and communicated if wind damage is to be avoided.

GUIDANCE

Slates are individually twice-nailed. Large slates (i.e. 500mm x 250mm and bigger) should not be head-nailed, especially in exposed locations; their increased leverage means they are more readily lifted by the wind, encouraging rain penetration, and progressively damaging the nail holes. Artificial slates are each fitted with an additional copper rivet and a retaining disk (set between the two slates below and turned-up to 'clip' their tails).

Double lap plain tiling is fully-nailed (two nails per tile) in exposed locations, at pitches over 60° or if nibless tiles are used. Otherwise, twice-nail every tile in every fourth or fifth course.

Single lap tiles only require fixing individually at pitches of 45° and over (at least one nail to each tile). The tail of each tile should also be mechanically fixed where the pitch is 55° or greater. Where separate "over and under" tiles are used (e.g. Italian or Spanish tiling), every tile should be nailed (two nails to under tiles) regardless of pitch.

For all types of roof, the end tile or slate in every course at verges or abutments, and at each side of valleys or hips should always be nailed, bedded in mortar or clipped, regardless of the type of tile, its lap or pitch. Likewise every eaves and ridge course (two courses for plain tiles), and all types of ridge and hip tile. Thick mortar bedding (e.g. to ridges over deeply profiled tiles) should be reinforced with small pieces of tile ('gallets') to preclude cracking and displacement. Hip irons should always be provided. Extra noggings will be required if dry-fixed systems are used with trussed rafters.

Care must be taken in the selection and setting out of slated and tiled roof coverings to avoid small, poorly-fixed (sometimes unfixed) cut pieces at junctions. One-and-a-half width slates and tiles must be used at verges and abutments.

Sheet metal coverings should incorporate clips within rolls, welts or standing seams, generally 50mm wide at 450mm minimum centres and nailed or screwed to the deck. The clips should be thicker in areas of severe exposure (e.g. 0.7mm rather than 0.6mm copper). Lapped joins (not drips) in roofs below 30° pitch and all free edges should be continuously clipped. Wood-cored rolls should be undercut on each side.

Single-lap tiles with a clipped verge

COMMON MISTAKES

- Nailing specification not provided by designers.

- Single-nailing of slates via central hole.

- Mortar bedding to hips and ridges too hard or too thick.

- Omission of hip irons.

- Small unfixed pieces of cut slate or tile at hips, valleys and abutments.

- Clips to supported metal roofing too far apart.

REFERENCES

BS 5534–1:1997 cl.3.2.6.5

BS 6915:1988 cl.5.2

BS CP 143–5:1964

BS CP 143–12:1970

BRE DAS 142

BRE Digest 311

BRE Report 302 pp.68 & 97

CDA TN 32 pp.10 & 30

LSA 1992 pp.16–25

ZDA 1988

Poor Detailing of Pitched Valleys to Slated and Tiled Roofs

THE PROBLEM   RATING **2 – 4**

The formation of a valley by the intersection of two roof pitches is a potentially significant weakness in the covering of a roof. Set at a pitch about 5° less than the roof, a valley is required to handle a significant concentration of water at a place where two different materials meet (i.e. the roof covering and a lining), or where the roofing material is used in a special way (e.g. swept valleys). Care must be taken in the detailing of valleys if they are to be effective in preventing water ingress.

GUIDANCE

Open valleys: The rafters each side are lined with 19mm (min. thickness) boards at least 225mm wide; these provide support for the two tilting fillets that run the length of the valley *and* the ends of the battens to the adjacent coverings. It is essential that the lining boards are set so that the top of the tilting fillets are level with the tops of the battens. For boarded or counter-battened roofs, the boards may sit on top of the rafters. Otherwise they are notched into the rafters (*not* trussed rafters) or supported on noggings and overlaid with 4mm ply (to ensure a smooth surface).

The width of the valley is determined by the location of the tilting fillets, which should be set to ensure that its depth is at least 75mm and its open width is not less than 125mm. Where the valley is more than 6 metres long and where the roof pitch is less than 35°, this width — and hence the width of the lining boards — should be increased. Calculations may be required for large roofs.

Linings are typically of sheet metal — usually lead — or GRP (not suitable for roof pitches lower than 22.5°). Code 4 or 5 lead linings are laid in lengths not exceeding 1.5m; heavier gauges can be up to 2m, provided the edges of the roof covering are not bedded. Laps must be in accordance with the following table (each length must be head-fixed with two rows of nails):—

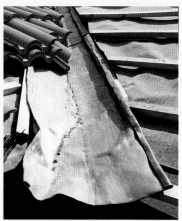

Poorly formed lead-lined open valley

Pitch of valley (5° less than roof)	15°	20°	30°	40°	50°	60°
Minimum length of lap (mm)	290	220	150	115	100	85

The cut (raked) edges of slates or double-lap tiles are laid dry to overhang the tilting fillet by at least 50mm. Single-lap tiles are bedded in mortar (a 25mm gap must be left between the bedding and the tilting fillet) which in turn is bedded on tile slips (or the lining is coated with bitumen paint).

Closed valleys (slates and double lap tiles only): Swept and laced valleys require a 25mm thick x 250mm valley board (fixed perpendicular to the valley rafter). Nailed to it are the curved battens. Purpose-made valley tiles and mitred valleys do not require additional support. A double layer of sarking felt should be laid beneath all closed valleys. Mitred valleys use lead soakers to seal the joint (code 3 or 4). They are suitable for valleys up to 6m long and roof pitches of 35° plus, although below 45° 'chevron' soakers must be used.

COMMON MISTAKES

- Lining boards not set flush with the tops of rafters (unless roof is boarded or counter-battened).

- Lining boards notched into trussed rafters.

- Lining boards not wide enough to support battens.

- Open width of lined valley less than 125mm.

- Sarking felt continued under linings to valleys.

- GRP liners used where rafter pitch is less than 22.5°.

- No slip plane (e.g. tile slips) between bedding and lining.

REFERENCES

BRE Report 302 pp.71 & 99
LSA 1990 pp.37–40
LSA 1991

Inadequate Specification of Metal Flashings

THE PROBLEM   RATING 3

Metal flashings are the first — and frequently the only — line of defence against water penetrating the horizontal and sloping junctions (abutments) formed where a roof meets a wall or a parapet. However, abutments are generally amongst the most exposed parts of a building which means that flashings have to be able to withstand extremes of wind, rain and temperature. Matters of size, configuration, support and fixing all have to be considered if water penetration at abutments is to be prevented.

GUIDANCE

Lead, copper and zinc can all be used to form flashings, though copper and zinc are not as well suited as lead to dressing over complex shapes such a bold-profiled tiles. The main types of flashing encountered on domestic roofs are:—

- **Horizontal cover flashings:** Simple strips of metal turned into a joint in the wall and overlapping the upstand of the adjacent roof by a minimum of 75mm. Flashings to kerb upstands (e.g. at the edge of a felt roof) must be supported and a building paper separator used to prevent the adhesion of the metal.

- **Stepped cover flashings:** Strips of metal at least 150mm wide are cut and formed into a rake of steps each of which is turned into a mortar joint. At least 65mm of metal is left to cover either a run of soakers (pieces of code 3 or 4 lead interleaved between slates or double lap plain tiles) or the upstand of a 'secret' gutter. The flashing can be extended by 150mm (200mm for low pitches in severe exposures) to cover single lap tiles, although secret gutters (75mm wide if open, 50mm if covered) should be used for flat-profile tiles.

- **Apron flashings:** Used where the head of a pitched roof abuts a wall. The flashing should have an upstand of at least 75mm and extend 150mm over the surface of the roof covering (200mm in severe exposures). Sheet metal must not be dressed or bossed over sharp-profiled tiles (to avoid thinning down and splitting); such flashings should be cut and welded.

Joints in flashings should be no further apart than 1.5m for lead (code 4 or 5), 1.8m for copper and 2.5m for zinc, and either lapped end-to-end (100mm for lead, 50mm for zinc) or single-lock welted (copper). Flashings must be turned-in to joints by a minimum of 25mm and secured by folded wedges at 450mm centres (600mm for zinc). For stepped flashings, every step must be wedged.

The free edges of all flashings must be secure. Lead and copper are held with clips at 300mm to 500mm centres, with thicker or continuous clips and extra fixings recommended for severe exposures. Zinc flashings 0.8mm thick and less than 100mm wide are stiffened with a half-bead or a 12mm welt. Otherwise, 40mm zinc clips at 600mm centres should be used. Lead cover flashings should also be side-clipped if their girth exceeds 150mm.

Flashing to a sloping abutment

COMMON MISTAKES

- Unsupported flashings over kerb upstands.

- Secret gutters not provided where flat, single lap tiles used at low pitches.

- Apron flashings dressed over tiles with sharp-profiles.

- Free edges of flashings not clipped (lead), or beaded or welted (copper and zinc).

- Wedging and fixing too shallow or infrequent.

- Mortar pointing to flashings too hard and brittle.

REFERENCES

BS 6915:1988 cls.5.4 & 6.9–6.10
BS CP 143–5:1964 cl.407
BS CP 143–12:1970 cl.4.11.3
BRE DAS 114
CDA TN 32 pp.10 & 30–31
LSA 1990 pp.6–19 & 63–67
ZDA 1988

Internal Walls and Floors

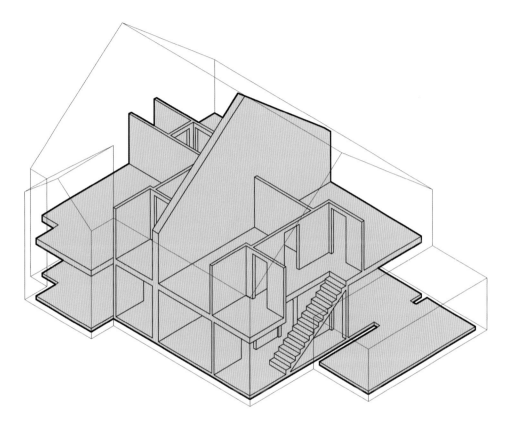

Defects and their Consequences

DEFECT	CONSEQUENCES			
	A	B	C	D
INADEQUATE MASS OF SEPARATING WALL OR FLOOR	3			
SOUND INSULATION REDUCED BY FLANKING CONSTRUCTION	4			
CAVITY OR FRAMED SEPARATING WALLS RIGIDLY CONNECTED	3			
LOCALISED OVER-STRESSING OF MASONRY PARTITIONS			3	
LACK OF STIFFENING TO JOISTED TIMBER FLOORS				3
FLOOR STRUCTURE WEAKENED BY SERVICES				2
INCORRECTLY SPECIFIED MDF STAIRCASES				2
FAILURE TO EFFECTIVELY CONNECT FLOORS TO MASONRY WALLS			2	3
LACK OF CONNECTION BETWEEN TIMBER JOISTS AND STEEL BEAM				3
BASE FOR TIMBER FLOORING INSUFFICIENTLY LEVEL				3
LACK OF SUPPORT TO TIMBER PANEL FLOORING				4
INCORRECT USE OF ABSORBENT OR RESILIENT MATERIALS	4			3
POOR DETAILING OF FLOATING FLOORS	4	3		4
INADEQUATE SPECIFICATION OF PLASTERBOARD LAYERS	3			2
INADEQUATE SPECIFICATION OF SCREEDS				3
INCORRECT THICKNESS OR GRADE OF TIMBER FLOORING				3
NO PROVISION FOR MOVEMENT IN TIMBER FLOORS				3
EXCESSIVE CUTTING OR CHASING INTO MASONRY WALLS			2	

The four columns labelled A to D refer to the four types of consequence detailed on the opposite page.

KEY TO HAPM RATING

1 Low probability of defect occurring, and only likely to have minor consequences.

2 Low probability of defect occurring, though with potentially serious consequences.
   or
   Reasonable probability of defect occurring, though only likely to have minor consequences.

3 Reasonable probability of defect occurring and with potentially serious consequences.
   or
   High probability of defect occurring, though only likely to have minor consequences.

4 High probability of defect occurring and with potentially serious consequences.

Consequences

A | INADEQUATE RESISTANCE TO THE PASSAGE OF SOUND

Current building legislation requires walls and floors between dwellings (and other parts of a building) to resist the passage of sound. However, the legislation does not actually state how much sound insulation is required (only in Scotland is there a 'deemed to satisfy' standard) and a variety of other legislation empowers local authorities to enforce sound insulation standards regardless. Nevertheless, it is generally accepted that one should not be able to hear normal conversation or a wc flushing in another property (airborne sound), nor footsteps from above (impact sound).

B | WATER STAINING, MOULD GROWTH AND FUNGAL DECAY

Moisture is always potentially harmful to buildings. Water penetration can stain, discolour and cause the deterioration of finishes. Condensation will encourage mould. Moisture can encourage fungal decay and insect attack, identification of which can be aided by the BRE's *Recognising wood rot and insect damage in buildings*.

C | DISTORTION AND CRACKING OF WALLS AND PARTITIONS

Separating walls and intermediate floors contribute significantly to the overall strength and stability of a building. Defects in their design can result in external walls leaning, bulging or buckling, and the formation of cracks, both at junctions between elements and through the elements themselves.

D | MOVEMENT, DEFLECTION AND DISTORTION OF FLOORS

Timber is a flexible, highly resilient material that is very sensitive to environmental change. Hence a timber floor or deck is naturally 'springy' and no floor can ever be anything like 'exactly' level; some opening-up of joints is also inevitable. Nor can a timber floor be expected never to squeak. Nevertheless, it is not generally acceptable if a floor springs so much that furniture 'shakes', or if the floor is so 'squeaky' as to cause persistent annoyance in a room below. Timber floors may also 'ridge' or 'arch' so badly that coverings − or even the deck itself − may fracture. Cracks may also appear in ceilings, and joints open-up enough for tongues to be disengaged.

Although concrete floors deflect only slightly, 'normal' deflection and irregularities can result in the cracking, curling and breaking-up of screeds and other finishes.

Inadequate Mass of Separating Wall or Floor

THE PROBLEM   RATING **3**

The mass of a masonry separating wall or a concrete separating floor determines how well it will resist the passage of airborne sound. On the whole, the greater the mass — which is measured in terms of area — the better. However, the acoustic performance of a wall or floor is also determined by the materials from which it is made, whether it incorporates a cavity, and how it is finished. Designers must take all of these factors into account and, if in doubt, err on the side of caution.

GUIDANCE

The minimum mass of a masonry separating wall should be:

Solid separating walls plastered or lined	375 kg/m² for brickwork	415 kg/m² [1] for concrete
Solid separating walls between isolated panels	300 kg/m² [2, 3] for brickwork or concrete	
Cavity separating walls 50mm+ cavity, plastered	415 kg/m² for brickwork or concrete	
Cavity separating walls 50mm+ cavity, lined	Not permitted in brickwork	415 kg/m² [4] for concrete
Cavity separating walls 75mm+ cavity, plastered or lined	Not permitted in brickwork	300 kg/m² [5] for concrete
Cavity separating walls between isolated panels	No limit [6] for brickwork or concrete	

[1] Insitu or panel walls must have a minimum density of 1500 kg/m³.
[2] 375 kg/m² if passing through a concrete separating floor.
[3] May be reduced to 160 kg/m² if concrete blocks with density of less than 1600 kg/m³.
[4] Mass of blockwork only, and only if stepped/staggered at least 300mm.
[5] Density less than 1600 kg/m³. Reduced to 250 kg/m² if stepped/staggered 300mm.
[6] 120 kg/m² each leaf if passing through a concrete separating floor.

Lightweight aerated concrete blockwork

COMMON MISTAKES

- Mass not clearly specified.
- Failure to appreciate that the strength of a block is independent of its mass.
- Lightweight blockwork not used in accordance with terms of BBA certificate.
- Floating screed assumed to contribute to mass (and hence acoustic performance) of floor.

Brickwork must be fully bonded and, in solid walls, concrete blocks must be full thickness.

Some types of lightweight aerated concrete blocks may also be acceptable. Their use should be justified by test data or by strict reference to a BBA certificate or similar third party guarantee.

The mass of a separating wall within a roofspace may be reduced to 150kg/m² if the ceiling below is of 12.5 mm plasterboard (or equivalent). Finishes may also be omitted, except where concrete blocks have a density of less than 1200kg/m³, in which case one side of the wall must be sealed.

Concrete floors should have a minimum mass of:

- 365 kg/m² if the floor is to be simply overlaid with carpet or some other form of soft covering.

- 300 kg/m² if the floor is to support a floating timber deck or screed (i.e. the deck or screed is separated from the concrete base by some form of resilient layer). Note that the 300kg/m² does *not* include the mass of the floating screed.

REFERENCES

BS 8233:1999
BRE Digests 333 & 334
BRE Report 238 (CIRIA Report 127)
BRE Report 358
HAPM TN 08

Sound Insulation Reduced by Flanking Construction

It is easy to forget that sound does not travel in a straight line, and that it can 'bypass' a separating wall or floor in a number of ways. The design of the construction 'flanking' a separating wall or floor is almost as important as the design of the wall or floor itself, and attention must be paid to factors such as the mass of external walls and partitions, the 'stopping' of cavities, and places where the flanking construction could reduce the mass of the wall or floor itself.

GUIDANCE

The minimum mass of a masonry flanking wall should be:

Solid external walls and the inner leaves of cavity external walls that abut timber separating floors	*375 kg/m² [1]*
Inner leaves of cavity external walls that flank separating walls or concrete separating floors	*120 kg/m² [2, 3]*
Partitions which pass through a separating floor	*375 kg/m² [4]*
Partitions that abut separating walls comprising a masonry core with independent linings	*Not permitted*

[1] Except where the wall has an independent lining.

[2] Unless the separating wall is solid and there are openings of at least one metre in height within 700mm of each side at every storey, or if it is of concrete block with a minimum mass of 415 kg/m³, in which case there is no requirement for a minimum mass.

[3] Except, in respect of a concrete separating floor, where at least 20% of the wall area of each room is taken up by openings and the conditions of note [2] are met.

[4] Masonry-cored partitions with an independent lining excluded.

Concrete ground and intermediate floors that cross timber framed separating walls must have a mass of at least 365 kg/m².

External wall cavities should be stopped with mineral wool or some other form of 'flexible' material (the practice of extending the separating wall to the outer leaf of the external wall is wholly unacceptable) and junctions between external and separating walls should be well bonded or tied (to ensure that gaps do not open up). All junctions with the timber framed walls should be sealed with tape or caulking and, if timber framed walls adjoin a timber separating floor, the edge of the floor should be blocked solid or otherwise stopped.

Services that penetrate the floor and pass though a habitable room should be enclosed above and below the floor. The enclosure should have a mass of at least 15 kg/m², and the duct should be lined or the services wrapped with 25mm unfaced mineral fibre; a sealed gap should also be left between the enclosure and the floor.

Look out for places where the flanking construction may reduce the mass of the separating floor or wall. For example, precast concrete planks are generally hollow and if built into a separating wall, their ends should be filled with concrete; beam and block floors also need to be built into the walls running parallel to the span (i.e. the mass of the floor must be maintained right through the supporting wall).

A flexible cavity closer

COMMON MISTAKES

• Lightweight blockwork not used in accordance with terms of BBA certificate.

• Separating wall extended across cavity to abut outer leaf of external wall.

• Inadequate size or position of openings.

• Built-in ends of hollow cored planks not specified to be filled.

• Edges of beam and block floors not built into flanking construction.

REFERENCES

BS 8233:1999
BRE Digests 333 & 334
BRE Report (CIRIA Report 127)
BRE Report 358
HAPM TN 08

Cavity or Framed Separating Walls Rigidly Connected

One of the key considerations in the design of both cavity masonry and timber framed separating walls is the degree to which it is possible to avoid connecting the two leaves of masonry or studwork. Ideally, there should be no connection, because sound is easily transmitted through rigid materials such as masonry, metal and wood. Any connection will reduce the effectiveness of the separating wall. However, separating walls are often 'structural' walls and hence connection may be necessary to ensure that the two leaves act as one. Connection may also be necessary to ensure the stability of the wall during construction. The problem for designers is how to connect the leaves without unduly compromising the acoustic performance of the wall.

GUIDANCE

In cavity masonry separating walls:

- The leaves must only be connected by 'butterfly' pattern wall ties (figure 1 to BS 1243: 1978), or certain types of proprietary tie that have been tested and shown to perform in a similar way, or which have been certified as being suitable by the BBA or some other third-party authority.

- The inner leaf of any external cavity wall must not connect the ends of the leaves, nor must the leaves of the separating wall be extended to meet the external outer leaf.

- Timber joists that bear on the leaves of the wall, must not be connected across the cavity (i.e. 'straddle' type joist hangers must not be used).

- Concrete floors that are built into the leaves of the wall must not cross, or even project into, the cavity.

In timber framed separating walls:

- The frames should be connected only by 40mm x 3mm metal straps fixed no closer than 1.2 metres, at (or just below) ceiling level.

- If it is necessary to firestop between frames, the material used must either be flexible (e.g. mineral wool) or fixed to one frame only.

If timber frames are separated by a masonry core, the frames should be at least 5mm clear of the core and only one frame connected to the masonry.

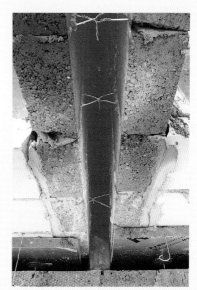

Butterfly pattern wall ties

COMMON MISTAKES

- Use of 'double triangle' or 'strip' type wall ties.

- Separating wall extended across cavity to abut outer leaf of external wall (i.e. ends of leaves rigidly connected).

- Cavity wall changed to solid above ceiling over top floor (i.e. tops of leaves rigidly connected).

- Masonry used to 'stop' cavity at roof level.

REFERENCES

BRE Report 238 (CIRIA Report 127)
BRE Report 358
HAPM TN 08

Localised Over-Stressing of Masonry Partitions

THE PROBLEM   RATING 3

Masonry partitions are often required to support lintels or beams in exactly the same way as external or separating walls. However, partitions are frequently considered to be 'non-load-bearing' and hence constructed of masonry that is less well bonded, thinner or weaker than the leaves of cavity walls. This can result in concentrations of stress that are actually greater than those that arise in walls which are seen as 'loadbearing' and, ultimately, the structural failure of the partition. It is therefore essential that lintel and beam bearings are designed to 'spread the load' and that the partitions as a whole are not subjected to loads for which they are unsuited.

GUIDANCE

Lintels that span doorways or other similar size openings will not usually cause any problems if each end has a minimum bearing of 150mm. However, this can present problems where openings of approximately the same height are closer together than 300mm. It may be better in these situations if a single lintel spans both openings, although a longer lintel may require a longer end bearing and care must be taken to ensure that no loads are carried by the slender pier between.

Bearings for beams should be designed by a structural engineer, who will take into account such factors as the height, crushing strength and tolerances of the supporting masonry. The layout and design of the partitions may have to accommodate padstones, spreader beams, bearing plates or additional piers, all of which must be clearly sized and specified.

Lintels or beams should never bear on short lengths of cut masonry, and the masonry at bearings must always be well bonded. The bonding pattern should ensure that there are no straight joints (i.e. perpends) immediately below or within close proximity to the ends of the bearings.

Where partitions are intended to be truly non-loadbearing, a gap should be left between the head of the partition and any floor above. Otherwise, the floor may bear down on the partition. The gap can be filled with a compressible, flexible material.

Masonry partitions must never be supported by timber floors.

Concrete lintel with insufficient bearing

COMMON MISTAKES

- Less than 150mm bearing to lintels and beams.
- Lintels or beams bearing on short lengths of cut masonry.
- Ends of beams located directly above perpends.
- No gap between head of non-loadbearing partition and floor over.

REFERENCES

BS 5628–3:1985 cls.19.1 & 19.3
BS 5977–2:1983 app.D1
BS 6178–1:1990 app.A & C
BS 8103–2:1996 cl.6.5
BRE GBG 21

Lack of Stiffening to Joisted Timber Floors

Sizes of timber floor joists in domestic-scale construction are usually selected from pre-prepared span tables, such as those contained in the various guidance documents that are issued in support of the Building Regulations (e.g. Approved Document A in England & Wales). However, these tables only ensure that a floor does not collapse. They do not take full account of the fact that timber floors − like all structures − deflect and distort under load, especially the sort of 'moving' loads to which floors are subjected (e.g. people moving around). For a joisted timber floor to be usable it must be stiffened.

GUIDANCE

Published span tables for timber joists are based on the assumption that the load on the floor will be evenly spread by the floor deck, which in effect means tongued and grooved boards or panels fixed to the top of every joist. For spans up to about 2.5 metres, a properly specified and fixed deck will provide adequate stiffness.

For spans greater than 2.5 metres, rows of herringbone strutting or solid noggings will have to be introduced at the following intervals:

Span (clear between supports)	Rows of strutting or noggins
2.5 to 4.5 metres	One at mid-span.
More than 4.5 metres	Two at the one third span positions.

Spaces between the joists and parallel walls should be packed solidly or wedged at the same intervals.

It should be noted that span tables do not generally cover floor joists which span more than about 5 to 5.5 metres. Floors that are required to span greater distances should be designed by a structural engineer.

Multiple joists (e.g. beneath partitions or around stair openings) can also deflect and distort unevenly if the joists are unable to act as one. Such joists must always be connected, either by bolts and toothed timber connectors spaced at approximately 1 metre centres, or by nails at a maximum of 450mm centres, about 20mm from both the top and bottom of the joists.

Triple joists bolted together

COMMON MISTAKES

• No packing or wedging to spaces at ends of lines of strutting or nogging.

• Double or triple joists not connected to act as one.

REFERENCES

BS 5268–2:1984
BS 6178–1:1990
BS 8103–3:1996 cls.6.0–6.33
BS 8201:1987 cls.6 & 26.2
TRADA WIS 1–36

Floor Structure Weakened by Services

THE PROBLEM RATING 3

It is quite common for the design of the heating, water supply, waste and electrical systems to be left for the plumber or the electrician to sort out on the basis of a 'performance' specification. While this approach is generally quite successful, it is sometimes the case that those responsible for the design of the services do not fully appreciate the implications of cutting into what may be structural elements. An ill-considered notch or hole can significantly weaken a floor. It is therefore important that the limits within which the plumber and electrician must work are made clear at the outset.

GUIDANCE

Notches in timber joists should *not* be:

- Deeper than an eighth of the depth of the joist.
- Further from the end of a joist than a quarter of its span.
- Closer to the end of the joist than seven hundredths of (0.07 times) its span (i.e. about 125mm for a joist spanning 1.8 metres and 350mm for one spanning 5 metres).
- Cut in both the top and bottom edge of the joist.

Holes in timber joists should *not* be:

- Larger than a quarter of the depth of the joist.
- Drilled through anywhere other than the 'neutral axis' which, for a joist of rectangular section, will be its mid-point.
- Closer to each other than three times their diameter, when measured centre to centre.
- Further away from the end of the joist than four tenths of (0.4 times) its span (i.e. 720mm for a joist spanning 1.8 metres and 2000mm for a joist spanning 5 metres).
- Closer to the end of a joist than a quarter of its span.

The above limitations apply only to simply supported joists. The limitation on cantilevered, propped and other types of joist should be determined by a structural engineer.

Services that pass through any type of concrete floor should be routed through specifically designed holes: holes must not be cut through concrete floors on an ad hoc basis, nor should individual blocks be randomly removed from beam and block floors. Holes through precast planks are factory-formed and must be designed at the outset.

Pipework notched to joists

COMMON MISTAKES

- Limits on formation of notches and holes not clearly stated.
- Holes for services in concrete floors not formally designed.

REFERENCES

BS 8103–3:1996 cl.6.2.2
BRE DAS 99

Incorrectly Specified MDF Staircases

THE PROBLEM   RATING 2

MDF — Medium Density Fibreboard — is regularly used in the construction of staircases. It machines well and requires little preparation prior to decoration. However, while MDF is a wood-based product, it differs from natural timber in two significant ways: it is weaker in tension and is unable to hold fixings nearly as well. This can cause problems for staircases, where the treads, risers and strings all rely on each other: closed treads gain strength and stiffness from being supported along their edges by the risers, which are themselves held in place by the treads; tight connections between the treads and the risers and the strings ensure that the weight of people ascending or descending the staircase is safely transferred to the adjacent walls and floors. Designers must therefore appreciate the peculiar properties and limitations of MDF.

GUIDANCE

For an 830mm staircase: treads should be at least 22.5mm thick with 9mm risers and 30mm strings (these figures are based on MDF with a minimum bending strength of $35N/mm^2$, which is considerably *above* the minimum performance standards for all types of board as defined in BS EN 622–5). The tread thickness will need to be increased for wider stairs, possibly the thickness of risers and strings too. Likewise winders (which always span further than the straight treads of a stair), unless metal angles or some other form of additional support is introduced. All loadbearing components should be of LA (dry conditions) or HLS (humid conditions) grade MDF to BS EN 622–5: 1997.

Treads and risers should be housed in rebates accurately machined into the strings. They must not be secured to the strings by way of wedges driven into trenched channels, as is customary with timber staircases. All joints must be glued and tightly screwed with parallel core screws (pilot holes will be required). Conventional wood screws and nails are unsuitable: they do not hold in MDF under load.

Screws should:

- be no further apart than 100mm and penetrate at least 19mm into the MDF base (i.e. the connection between a 9mm riser and the back of a 22.5mm tread would require a screw of at least 28mm in length).

- be at least 12mm in from the edges and 25mm from the corners of components.

- not penetrate the edges of boards within 70mm of the corners.

This will almost invariably mean that the thickness of some components may have to be increased to ensure a close fit.

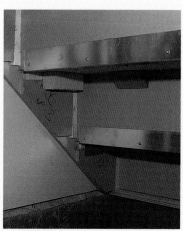

Strengthening a defective MDF stair

COMMON MISTAKES

- Winders too thin for span.
- Treads and risers trenched and wedged to strings.
- Specification not explicit in requiring use of parallel core screws.
- Fixings too close to edges of components.

REFERENCES

BS 585–2:1985 cls.4.2 & 6.2–6.6

BS 7916:1998 cls.11–13

BRE Digest 435

FIDOR SR6

FIDOR X509

Failure to Effectively Connect Floors to Masonry Walls

THE PROBLEM   RATING 2 – 3

A masonry wall which is too thin for its height will buckle and become unstable, regardless of its ability to support any vertical load. Walls may also be exposed to 'eccentric' forces set up by the elements, especially the push and pull of the wind. However, increasing the thickness of a wall to suit its height is only practical within certain limits: stability has to be provided in some other way. This means designing the walls and floors to act together: the walls support the loads imposed by the floors which, in turn, counter their tendency to bulge or lean. It is therefore vital that the connection between floors and walls can transfer tensile and compressive forces.

GUIDANCE

For concrete floors, an effective connection can be achieved by simply building them into the walls by at least 100mm although, due to their camber, long planks may need to be dry packed. The one exception is the 'parallel' edge of a 'beam-and-block' floor (see below).

In buildings up to two storeys high, built-in timber joists will also suffice, provided they sit on at least 90mm of masonry or a 75mm wall plate, and are at 1.2m maximum centres.

While there are many ways in which the junction between a timber-joisted floor and a supporting wall could be designed, the most common is to use 'restraint' straps – strips of galvanised or stainless steel (min. 5mm x 30mm x 1200mm long with a 100mm 'hooked' end) screwed or nailed to the joists and built into the wall. These should be at 2m maximum centres, 1.25m if the building has more than three storeys. Greater centres are acceptable if the edge of the floor is interrupted by an opening (e.g. a stairwell), provided the opening is no longer than 3m and the total number of straps is not reduced. Noggings are needed between the ends of the joists (to stop them twisting).

Straps can also be used at right angles to both concrete beam-and-block and timber joisted floors.

Concrete floors: straps should be long enough to connect to the first two beams, although no shorter than 800mm.

Timber floors: straps should be long enough to connect to the first three joists and supported by 38mm thick noggings (half the depth of the joist for straps fixed to the tops of the joists, full depth if fixed to the bottoms) fixed tight between the joists, and packed between the end joist and the wall.

Where there is a floor each side of a wall (e.g. at a separating wall), the floor need not be connected to the wall provided the floors are at about the same level, in continuous contact with the wall or, at the very least, in contact at no more than 2m intervals (in plan).

Restraint strap at 90° to timber joists

COMMON MISTAKES

- Failure to account for camber of precast concrete planks.

- No straps provided to beam and block floors.

- Straps face-fixed to walls internally.

- Straps hooked over cavity wall insulation rather than face of masonry.

- No packing or wedging of gap between wall and first timber joist (for straps at 90°).

REFERENCES

BS 6178–1:1990 app.E

BS 8103–1:1995 cl.5.4

BRE DAS 25

BRE GBG 21 & 29–1

TRADA WIS 1–36

Lack of Connection between Timber Joists and Steel Beams

THE PROBLEM   RATING 3

Steel beams are often used in house building. They give designers the freedom to locate partitions at will, while maintaining the practicality of timber joists. Sometimes, the joists are nailed to timber plates bolted to the steelwork, which gives a secure, 'conventional' connection. However, in other cases, they are supported by metal joist hangers set on the beams which can present problems: there is unlikely to be any surface against which to set the backplate of the hanger (rolled hollow sections or channels would be exceptions) and, if the beam does not support a masonry wall, there will be nothing to hold the hanger in place. It is therefore important that designers select the right type of hanger for each situation and consider fully how it is connected to the beam.

GUIDANCE

If the steel beam supports a bonded masonry wall the joists may be supported on ordinary – 'standard' or 'restraint' type – masonry hangers provided that the height of the masonry above the hanger is not less than the minimum specified by the hanger manufacturer. The space between the back of the hanger and the face of the beam should be packed tight, especially if the depth of the joist hanger is less than that of the beam.

Where beams do not support any masonry, or the height of any masonry is less than as recommended by the hanger manufacturer, ordinary masonry hangers cannot be used. However, some types of 'stirrup' hanger may be used if:–

- The width of the stirrup fits exactly the width of the beam.

- Joists are supported in pairs, one each side of the beam.

- The loads each side of the beam are approximately equal.

- All spaces behind the hangers are tightly packed.

Ideally, no hanger should be deeper than the beam, but, if 'underslinging' is unavoidable, the distance between the underside of the beam and the hanger should not exceed:

- 30mm for a standard or restraint type hanger.

- Half the depth of the hanger for the straddle variety.

These distances may be exceeded on the advice of a structural engineer. Always check with the manufacturer of the hanger that it is designed to be 'underslung' and – again – ensure that the spaces behind all parts of the hangers are tightly packed.

Timber joists and packings will almost inevitably shrink in service, and the detailing of the connection between the joists and the beam must take this into account. The tops of the joists should be at least 12mm above the top of the beam and all timber packing should be fixed in position to prevent it becoming dislodged.

Timber joists bearing on a steel beam

COMMON MISTAKES

- Insufficient weight of wall above ordinary masonry hangers ('standard' or 'restraint' type).

- Spaces between hangers and steel beam or between hangers that are 'back to back' not packed tight.

- No allowance for potential shrinkage of joists or timber packings.

REFERENCES

BS 6178–1:1990 app.E
BRE GBG 21

Base for Timber Flooring Insufficiently Level

THE PROBLEM
RATING **3**

A great deal of effort goes into selecting timber board and panel products for use as flooring, and to ensuring that they are well-laid and finished. However, it is often forgotten that the final quality of a floor is heavily dependent on the quality of the structure on which it is laid, something which is as true for a floor which 'floats' on insulation or battens as it is for one which is fixed direct to timber joists. Variations in level between the tops of joists can result in an undulating floor with raised edges and fixings. Irregularities in a concrete base can mean that non-structural battens or insulation are forced to 'span' between high points in a way which was never intended. Designers should be clear as to how level the base must be and, if necessary, how this may be achieved.

GUIDANCE

A timber joisted floor will be level enough for ordinary board or sheet flooring if the joists are specified to be 'regularised' (i.e. the machining or sawing of every piece of timber in a batch to a uniform size), or of 'surfaced' American or Canadian lumber ('ALS' or 'CLS' timber).

Account should also be taken of the likely shrinkage of the joists as the finished building 'dries out': a clearance of at least 12mm should be left between the underside of the floor deck and the top of any beam or partition below (so as the beam or partition does not 'push up' against the deck), and multiple joists should be bolted together using toothed connectors (so they act as one and do not twist or shrink differentially).

The surface of a concrete base which is to support a floating floor should not deviate more than 5mm under a 3m straight edge. Nor should there be any abrupt changes of level, such as may occur between the joists and infill of a 'beam and block' floor. This may mean that the surface will have to be levelled with either a thin layer of mortar (1:6 cement:sand should suffice), or some form of proprietary levelling screed. Sand must never be used, due to its tendency to move around ('pump') when the floor is subject to traffic.

Battens for floating floors should be both regularised and deeper than any insulation between. Some proprietary systems require the battens to sit on purpose designed cradles which require special levelling. Manufacturer's instructions should always be followed.

Beam-and-block floor base as laid

COMMON MISTAKES

- Timber joists or battens for flooring not regularised.

- Lack of clearance between tops of timber joists and partitions or beams below.

- Abrupt changes in level of surfaces of beam and block floors.

- Sand used to level concrete floor bases.

REFERENCES

BS 7916:1998 cl.5.4.2

BS 8201:1987 cl.6.4.2

BS EN 1313–1:1997

BRE GBG 28–1

BRE Report 262 pp.47 & 48

TRADA WIS 1–36

Lack of Support to Timber Panel Flooring

THE PROBLEM   RATING **4**

Floors made up of large panels of plywood, particleboard or oriented strand board (OSB) are commonly used in lieu of narrow timber boards or strips. Their large size (typically 2400mm x 600mm or 1200mm) means floors can be laid more quickly and cheaply, and with fewer joints. However, traditional timber boards and strips (65mm to 137mm wide) do have one distinct advantage: their long edges are not subject to anything like the stresses experienced by the edges of large panels, since floor loads are almost always distributed across a number of boards and hence via a number of joints. This is not the case with panel flooring, where it is perfectly possible for the same load to bear on a single panel, and thence via a single edge, a problem which may be exacerbated by the fact that timber-based panels are stronger in one direction (the longer) than the other. It is therefore important that designers are aware of the structural limitations of panel-based flooring materials, and that their instructions for the laying of the boards and the provision of any necessary additional support are clear.

GUIDANCE

All types of tongued and grooved flooring panels should be laid with their long edges running at right angles to the joists or battens. OSB panels with square edges must be laid likewise (i.e. the short edges of OSB panels should not span supports).

Square-edged plywood or particleboard panels can be laid either direction, although they are best laid with their long edges continuously supported along the tops of the joists or battens. This means that the joist/batten spacing may have to be adjusted to suit the widths of the panels.

Support for T&G floor panels

It should be noted that square edged boards should not be used in the construction of a separating floor.

Additional support – in the form of noggings or counter battens – should be provided:

- to the edges of any panels which are not continuously supported within 50mm of the face of a wall.

- to the short edges of any tongued and grooved panels which are not continuously supported by a joist or batten.

- to any square edges which are not continuously supported by a joist or batten.

- where the continuity of the floor or its support is interrupted by a fire place or any other rigid upstand.

- around the edges of access panels.

Timber panels in floating floors also need to be supported where any extra loading is anticipated, such as beneath the edges of kitchen fittings and appliances, and at door thresholds.

COMMON MISTAKES

- Edges of flooring panels not supported where last joist or batten further than 50mm away from wall.

- No additional support to short edges of tongued and grooved boards or edges of square edged boards that occur between joists or battens.

REFERENCES

BS 7916:1998 cls.5.31, 5.33 & 5.4

BS 8201:1987 cls.5.3.1 & 26.2–3

BS EN 300:1997

BRE Digest 323

BRE DAS 31

HAPM TN 11

TRADA WIS 1–28 & 1–36

WPIF 1998

Incorrect Use of Absorbent or Resilient Materials

THE PROBLEM   RATING **3 – 4**

Absorbent and resilient materials (e.g. mineral fibre, expanded polystyrene, etc.) are regularly used to improve the sound resistance of certain types of separating wall and floor. A 'quilt' or a 'curtain' within a timber-framed separating wall or a suspended timber separating floor will absorb sound, enhancing the ability of the wall or floor to resist the passage of airborne sound. Resilient 'layers' or 'strips' are used to isolate screeds or timber rafts from the floor base, lending the floor its ability to resist the passage of impact sound. The absorbent/resilient properties of a material depend on its thickness and density. Both must be specified: a material which is too thin or of insufficient density will have a light, open texture and little ability to absorb sound and, if used as a resilient layer, will compress under load and fail to isolate the floor from its base.

GUIDANCE

Absorbent or resilient materials for use in sound resisting construction should comply with the following table:–

Absorbent curtain within timber framed separating wall	25mm unfaced mineral fibre with a density of at least 10kg/m³ [1]
Absorbent blanket between joists of timber suspended floor	100mm unfaced mineral wool or rock fibre with a density of at least 10kg/m³ at ceiling level [2]
Resilient layer between a concrete base and a screed	25mm mineral fibre with a density of at least 36kg/m³
	13mm of pre-compressed expanded polystyrene board [3]
	5mm of extruded closed-cell foam with a density of between 35–45 kg/m³
Resilient layer between a concrete base and a timber raft (timber or timber-based boarding on 45 x 45mm battens)	25mm mineral fibre with a density of at least 36kg/m³ [4]
Resilient layer between a timber floor base and a floating layer	25mm mineral fibre with a density of between 60–100kg/m³
Resilient strips between the tops of joists and a ribbed floor (45mm x 45mm battens fixed to underside of timber or timber-based deck)	25mm mineral fibre with a density of between 80–140 kg/m³

Fibreglass quilt used as a resilient layer

COMMON MISTAKES

- **Density of mineral fibre products too low.**
- **Use of proprietary products not backed by testing or third party certification.**

[1] 50mm if attached to one of the frames, or two 25mm layers, one attached to each.

[2] For ribbed floors, pugging with a mass of at least 80kg/m² may be used in lieu.

[3] To be laid with lapped joints.

[4] May be reduced to 13mm if the battens incorporate an integral closed-cell foam strip.

Proprietary products may also be used, provided their performance is backed by a BBA Certificate or other recognised third party certification, or has been demonstrated by tests.

REFERENCES

BRE Digests 333 & 334
BRE Report 238 (CIRIA Report 127)
BRE Report 358
HAPM TN 08

Poor Detailing of Floating Floors

THE PROBLEM   RATING **4**

A floating layer can enable an insitu or precast concrete floor to achieve a good standard of thermal or acoustic performance, while providing the option of using a variety of floorings. However, the benefits of using a floating floor will be negated if the detailing of the floor does not prevent the 'bridging' of the resilient or insulating layer, its integrity is compromised by the ill-considered placing of services, or — in the case of floors that include moisture sensitive materials — it is not protected from damp; a failure to consider any of these factors can result in a floor which falls well short of its desired performance.

GUIDANCE

Floating layers should be turned-up to isolate the edges of the flooring (a 10mm gap – stopped with resilient material – should be left between the boarding/screed and surrounding walls). If the floating layer serves to provide sound insulation, it should not be bridged by skirtings or services. Skirtings should be fixed to the walls with a 3mm gap left between the skirting and the floor. In timber floors, the gap should be sealed with acrylic or neoprene (to prevent moist air condensing within the floor).

For screeded floors, there must be no gaps in the floating layer through which the screed can migrate. Joints between semi-rigid or compressed materials (e.g. extruded polystyrene) should be rebated or taped. Material with an 'open' structure (e.g. mineral fibre) should be paper-faced or overlaid with a membrane.

Ideally, horizontal services should not be run within the depth of a floating floor. If this is not possible, they should be run in purpose made ducts, although this too can be a problem:

- Resilient layers used for sound insulation are likely to be too compressible to adequately support a duct, although some types of closed-cell foam and certain proprietary products may be suitable, provided the duct is held in place by a well compacted screed or battens fixed to a timber deck (i.e. laid loose on the resilient layer).

- Ducts within floating floors can result in cold bridging. Either the insulation must be able to support a duct within the depth of the flooring or the duct itself must be insulated (e.g. all voids in the duct filled with mineral wool).

All floating floors should incorporate a vapour control layer (e.g. 500 gauge polythene) unless there is already a DPM *on top* of the concrete base. This layer – which protects timber or moisture sensitive finishes from construction moisture – should sit above the insulation (unless pre-bonded panels are used) and be turned-up at the edges and sealed below the skirting.

Service duct within a floating floor

COMMON MISTAKES

- Edge gaps not stopped.
- No gap between skirtings and surface of floor.
- Gap between skirtings and timber floor not sealed.
- Joints in insulation boards beneath screeds not rebated or taped.
- Ducts fixed through resilient layers (may increase sound transmission).
- No vapour control layer.

REFERENCES

BS 8233:1987 cl.14.5.5
BRE Digest 334
BRE Report 262 pp.47–50
BRE Report 358
HAPM TN 11
WPIF 1998

Inadequate Specification of Plasterboard Layers

THE PROBLEM   RATING 2 – 4

Layers of plasterboard are an essential component in a number of types of separating floor and wall. Used as wall linings, they provide the mass in a timber framed separating wall; as isolated panels each side of a masonry separating wall they permit the use of masonry of a lower mass than would normally be the case. Plasterboard can also provide the greater part of the mass in a timber separating floor, and may also be used as a wall lining (in lieu of wet plaster). The thickness and number of layers of plasterboard can be critical, as can the nature of their installation.

GUIDANCE

Timber framed separating walls should be lined each side with two layers of plasterboard with staggered joints and a total combined thickness of no less than 30mm. The linings should be at least 200mm apart. Power points and other services should be backed-up with cladding of the same thickness, except in Scotland or Northern Ireland where services within timber framed separating walls are not permitted.

For separating walls that consist of a masonry core between isolated panels, each lining (panel) should comprise:

- Two sheets of plasterboard joined by a cellular core with a mass (including any plaster) of at least 18 kg/m^2, or

- Two sheets of plasterboard with staggered joints and a combined thickness of at least 30mm (25mm if a supporting framework is used).

Joints between panels should be taped and the panels (or their supports) fixed to the ceiling or floor only: on no account must the panels be fixed to the masonry core. An airspace of at least 25mm should be left between the panels and the masonry core.

Linings to masonry separating walls should be 12.5mm thick.

Ceilings to timber separating floors should comprise:

- Two sheets of plasterboard with staggered joints and a total thickness of not less than 30mm, or

- For floors with pugging, two sheets of plasterboard with staggered joints and a total thickness of not less than 25mm with a 6mm sheet of plywood fixed between the plasterboard and the joists (19mm of dense aggregate plaster on expanded metal lath can be used in lieu).

Noggings should be provided to support the perimeters of all ceilings. Ceilings which support pugging should be capable of supporting the additional weight with a plastic sheet between the ceiling and the joists.

Edge supports to a plasterboard ceiling

COMMON MISTAKES

- Integrity of plasterboard to timber framed separating walls compromised by services.

- Inadequate thickness of linings to masonry walls.

- No support to perimeters of ceilings.

REFERENCES

BS 8000–8:1994 cl.3.1.8
BS 8212:1995 cls.3.2 & 6.1
BRE DAS 81
HAPM TN 05

Inadequate Specification of Screeds

THE PROBLEM   RATING **3**

The purpose of a screed is to provide a smooth and level base for the laying of floorings, and hence most screeds used in domestic scale construction contain only cement and sand. However, the absence of any coarse aggregate means that it is difficult to control drying shrinkage, and that the screed may curl or crack as a result. It is with this possibility in mind that sand and cement screeds are mixed using the minimum of water, are covered while curing and then allowed to dry slowly (i.e. heating should not be brought into service too quickly and dehumidification should not be used). However, these measures alone cannot control shrinkage: designers must consider the mix and thickness of screed, whether it needs to be reinforced, and the nature of the base on which it is to be laid.

GUIDANCE

Cement and sand screeds should have a mix proportion of between 1:3 and 1:4 cement:sand measured by mass. It is preferable that screeds over 50mm in thickness are of fine concrete with a mix of between 1:4 and 1:5 cement:aggregate measured by mass.

- **Fully bonded screeds** (i.e. those laid on a set and hardened base which has been scabbled or otherwise roughened) should have a minimum thickness of 25mm and a maximum thickness of 40mm.

- **Unbonded screeds** (i.e. those laid on a damp proof membrane or other separating layer, or those laid on a set and hardened base with only a tamped surface) should have a minimum thickness of 50mm.

- **Floating screeds** (i.e. those laid on a layer of insulation or similar compressible material) should have a minimum thickness of 75mm, although a minimum thickness of 65mm is adequate for domestic or other lightly loaded situations.

Freshly laid cement:sand screed

COMMON MISTAKES

The thickness of a screed should be measured at its thinnest point, accounting for any camber or other variation in the base.

Unbonded and floating screeds should be reinforced, either with a layer of steel mesh (steel fabric reference D49 or D98 to BS 4483 is considered acceptable), chopped fibre or some other proprietary material. Local reinforcement should be provided over pipes or additional insulation.

Anhydrite and semi-hydrate screeds may be laid thinner than cement and sand screeds. Lightweight screeds will require a thin, dense topping. In both cases, proprietary products are normally used; manufacturer's instructions must be followed.

- Thickness of screed not measured at its thinnest point (i.e. no account taken of cambers or irregularities in base).

- Fully bonded screeds too thick.

- Omission of reinforcement to unbonded screeds.

- No additional reinforcement over services.

REFERENCES

BS 8204–1:1987 cls.5.4.3 & 6.4.1
BRE Report 332 pp.157–176

Incorrect Thickness or Grade of Timber Flooring

THE PROBLEM   RATING 3

A wide variety of timber and timber-based products are used for flooring in housing, including softwood boards or strips, plywood sheets, and panels of particleboard (chipboard), cement bonded particleboard or oriented strand board (OSB). All have their uses and all have their own distinct properties which designers must consider when determining the thickness and grade of material to be used in any particular location.

GUIDANCE

Softwood tongued and grooved strips and boards should comply with BS 1297:1987 and the following table:

Finished Thickness	16mm	19mm	21mm	28mm
Maximum Span	505mm	600mm	635mm	790mm

The following types of panel are suitable for use in housing:

- Plywood to BS EN 636-1:1997 (dry conditions) or BS EN 636-2:1997 (humid conditions)
- Particleboard grade P5 or P7 to BS EN 309:1992
- Cement bonded particleboard to BS EN 633:1994
- Oriented Strand Board grade 3 or 4 to BS EN 300:1997.

Grade OSB/2 can also be used as long as there is no risk of it becoming wet. It should be noted that no type of particleboard or OSB can ever be 100% 'moisture resistant' and if the floor will be subject to anything other than occasional wetting, such boards should not be used. Particleboard and OSB floors in kitchens, bathrooms and other potentially wet areas should be overlaid with a vinyl sheet covering with welded joints and a 'coved' skirting.

Particleboard after prolonged wetting

COMMON MISTAKES

- Use of incorrect grade of particleboard.
- Failure to note limitations of 'moisture resistant' boards.
- Particleboard or OSB floors in kitchens or bathrooms left uncovered in service.

The thickness of timber-based panels should be as follows:

Span between supports	450mm	600mm
Plywood	12mm	16mm
Particleboard	18/19mm	22mm
Cement Bonded Particleboard	18/19mm	22mm
OSB Grade 2	18/19mm	22mm
OSB Grades 3 & 4	15mm	18/19mm

OSB Panels must be laid with their 'major axis' (marked on each board) spanning between supports.

For continuously supported ('floating') floors, the thickness of the flooring should relate to the stiffness of the floating/resilient layer, although in no case should it be less than 18mm. OSB should never be used.

REFERENCES

BS 8201:1987 cls.10.1 & 26.3
BS 7916:1998 cls.5.3.1 & 5.3.3
BRE Report 332 pp. 277–281
HAPM TN 11
WPIF 1998

No Provision for Movement in Timber Floors

THE PROBLEM   RATING 2

Timber is a hygroscopic material: it reacts to changes in temperature and humidity by releasing and absorbing moisture, shrinking and swelling as a result. Depending on species, a piece of softwood can be expected to shrink or swell in section (i.e. across its grain) by about 0.3% for every 1% change in moisture content. Particleboards shrink or swell equally in all directions and can be delivered to site with a moisture content of as low as 2%: a 2.4 metre board could expand by up to about 4mm when introduced into an internal environment. Designers must be clear in stating under what conditions the floors are to be laid and what measures are needed to accommodate movement.

GUIDANCE

Boards or panels should be laid with a moisture content of:–

- 15–19% for unheated areas (max. 15% for particleboards).
- 10–14% for intermittent heating (9–12% for particleboards).
- 9–11% for continuous heating (7–9% for particleboards).

These figures are drawn from BS 8201 and BS 7916. However, moisture conditions in service may be much lower. It may therefore be prudent to use the figures set out in BS EN 942:–

- 12–16% for unheated buildings.
- 9–13% for buildings heated to 12–21° C.
- 6–10% for buildings heated to over 21° C.

Movement in service can be minimised by ensuring that:–

- A 10–12mm gap is provided around all the edges of all floors. Particleboard and OSB may need wider or additional gaps (to allow for an expansion of 2mm per metre).
- Floating floors are divided into room-sized bays.
- Floors on concrete bases are not laid until the concrete is dry, or the flooring is protected by a damp proof membrane.
- Timber joists and battens have a moisture content of more than 20% at the time the floor is laid.
- Narrow strips or boards are used in preference to wide, and the strips/boards are pulled together ('cramped') before fixing.
- Tongued and grooved joints in panel flooring are continuously glued and wedged together tight.
- Particleboards and OSB are not laid until plastering and other wet site operations are complete.
- Heating pipes within or below floating floors are insulated.

Excessive shrinkage gaps in T&G flooring

COMMON MISTAKES

- Inadequate or inappropriate specification of conditions under which flooring to be installed.
- Floating floors not divided into room-sized bays (i.e. floor constrained under non–loadbearing partitions).
- Tongued and grooved joints in panel flooring not glued or only glued intermittently.
- Plastering carried out after laying of particleboard or OSB panels.

REFERENCES

BS 7916:1998 cls.5.4 & 5.5
BS 8201:1987 cls.6.9.2 & 28.3.3
BRE Digest 334
BRE Report 332 pp.279 & 281
HAPM TN11
TRADA WIS 1–36 & 4–27
WPIF 1998

Excessive Cutting or Chasing into Masonry Walls

THE PROBLEM   RATING 2

It is not generally possible in simple masonry construction to avoid the need to cut or chase for electrical (and sometimes other) services. Electrical boxes and 40+ amp wiring (as used for cookers) are considerably thicker than the usual 13–15mm of plaster or plasterboard lining, and even 5 amp wiring (as used for lighting) will, when covered, unacceptably reduce the thickness of finishes. Most cutting and chasing is formed on site and if not carried out with care can impinge on the structural performance of the masonry and — if in the context of a separating wall — its acoustic performance. It is therefore important that designers make clear the limits on the cutting and chasing of masonry at the outset.

GUIDANCE

From a structural point of view, chases should be formed and positioned in accordance with the following criteria:

- Vertical chases should not be deeper than 1/3 of the wall thickness or – in cavity walls – 1/3 of the thickness of either leaf.

- Horizontal chases should not be deeper than 1/6 of the wall thickness.

- Where hollow blocks are used, 15mm of block material should be retained between the void and the chase.

In terms of acoustic performance, it is preferable that there are no chases or recesses in a masonry separating wall. However, there will be situations when this is not possible, in which cases the chases or recesses on each side of the wall should not be located 'back-to-back'.

Note that the BBA certificates for some types of lightweight aerated concrete block also limit the horizontal proximity of chases and recesses each side of a separating wall.

Chases in masonry separating walls should always be well filled with mortar after wiring is complete.

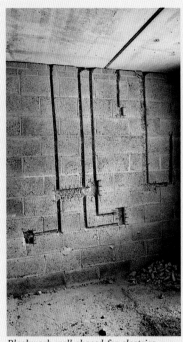

Blockwork wall chased for electrics

COMMON MISTAKES

- Chases cut too deep into hollow blocks.

- Recesses or chases set 'back to back'.

REFERENCES

BS 5234–1:1992 cl.3.6.2
BS 5628–3:1985 cl.19.6
BS 8103–2:1996 cl.6.9

Above Ground Services

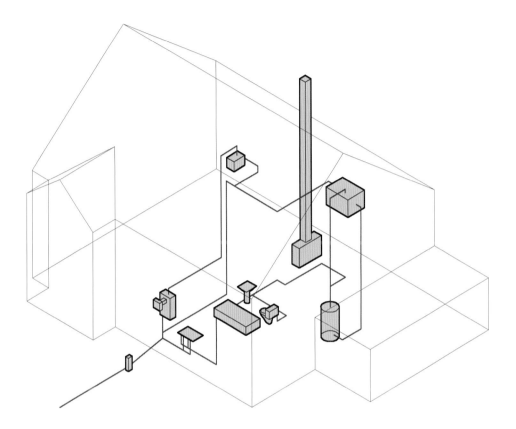

Defects and their Consequences

DEFECT	CONSEQUENCES			
	A	B	C	D
BASIC DESIGN OF SERVICES NOT CONSIDERED AT THE OUTSET	2	3	1	1
INADEQUATE DESIGN OF OPEN FLUES		2		4
PLANT OR COMPONENTS UNSUITED TO WATER SUPPLY	3	3	1	
BADLY PLANNED VENTILATION SYSTEMS			3	2
INADEQUATE DESIGN OF WASTE OR RAINWATER DISPOSAL SYSTEMS			2	
INSUFFICIENT SUPPORT TO PIPES OR GUTTERS			2	
WALLS, FLOORS OR ROOFS NOT DESIGNED TO SUPPORT SERVICES			1	1
FAILURE TO TAKE ACCOUNT OF OPERATING ENVIRONMENT	2	2		
LACK OF SYSTEM PROTECTION	2	3		3
NO FORMAL PROVISION FOR COMMISSIONING AND TESTING	3	4	3	3

The four columns labelled A to D refer to the four types of consequence detailed on the opposite page.

KEY TO HAPM RATING

1 Low probability of defect occurring, and only likely to have minor consequences.

2 Low probability of defect occurring, though with potentially serious consequences.
or
Reasonable probability of defect occurring, though only likely to have minor consequences.

3 Reasonable probability of defect occurring and with potentially serious consequences.
or
High probability of defect occurring, though only likely to have minor consequences.

4 High probability of defect occurring and with potentially serious consequences.

Consequences

A ⬛ TOTAL FAILURE OF SYSTEMS OR COMPONENTS

Defects in the design or specification of services can seriously reduce the reliability of components and systems, and lead to the total failure of an installation to perform its intended function. For example, overheating may cause components to crack or distort, and blockages can prevent the flow of liquids or gases. In some circumstances, there may be no obvious signs of damage other than the fact that something does not work. Note: The total failure of a system or component should not be confused with the inability of an installation to achieve "fitness for purpose". A heating system that does not provide heat has suffered a total failure; one that is not capable of achieving a desired internal temperature is merely not "fit for its purpose", something that is not the concern of this guide.

B ⬛ PREVENTION OF OTHER COMPONENTS FUNCTIONING

Services components are dynamic and interactive, both in relation to the installations of which they form a part and – in more general terms – the fabric of the buildings they serve. There will be many situations where the performance of a component is heavily dependent on the performance of another, possibly located some distance away and not obviously part of the same system (e.g. a pressure-reducing valve that regulates the water supply to a directly fed water heating appliance). Even systems which appear to be totally independent can interact in unforeseen ways, such as when one may have an adverse influence on the operating environment of another.

C ⬛ WATER STAINING, MOULD GROWTH AND FUNGAL DECAY

Many services installations carry or store water, providing the opportunity for the leakage of both liquids and vapour due to the corrosion of metals, the breakdown of plastics or the failure of joints. Any form of moisture is potentially harmful to buildings, its presence being indicated by staining, discoloration, mould growth and – in extreme cases – the onset of fungal decay and insect attack (for identification, reference should be made to the BRE publication *Recognising wood rot and insect damage in buildings*).

D ⬛ DIRECT THREAT TO HEALTH AND SAFETY

DANGER OF DEATH

Mechanical and electrical services systems have the potential to pose a serious threat to the health and safety of persons who occupy or maintain buildings, even simple housing. High temperature water within heating and hot water systems can scald and burn, many gases are lethal asphixiants (especially the products of combustion) and electric shocks can kill. Poor quality water – possibly the result of contamination by sewage or active bacteria – can cause disease and serious long term illness.

Basic Design of Services not Considered at the Outset

Simple mechanical and electrical services – as generally found in single family houses – can be (and often are) successfully designed using little more than widely accepted 'rules of thumb'. However, this does not mean that designers are absolved of all responsibility for their installation. Many basic design decisions still need to be made, without which the plumber or electrician will not know where to begin. Worse still, failure to consider certain fundamental issues can actually inhibit the installation of the services, creating 'impossible' situations (e.g. sanitary fittings that are difficult to drain or tanks that cannot be installed high enough) and forcing defective design. It must also be recognised that it does not take much for a 'simple' domestic services installation to become a complex one (e.g. in an old persons home or a building in multiple occupation), and for the ad hoc approach to the design of such a system to create its own problems. The design of all building services must be considered at the outset of a project if installation is to progress smoothly and systems are to function as intended.

GUIDANCE

A design brief or performance specification should be prepared at the outset for all types of services installation. This can then be used to ensure that the basic points of the design are given due consideration, and as a tool to determine whether formal design by an independent services consultant is in any way required.

Amongst the basic points that should be considered in the design of any services installation are:-

- The sizes of major items of plant or equipment, space for which must be allocated at the planning stage (including for future maintenance and replacement). Many problems can be avoided if feed cisterns are sited above storage cylinders which in turn are sited above boilers, and that equipment is positioned to permit flues, pipes and ducts to be straight, vertical or (if appropriate) horizontal.

- The layout of flues, ducts and soil stacks should avoid conflict with structural elements and permit the accurate formation of holes, chases, etc. in prefabricated elements such as wide span concrete plank floors.

- The weight and fixing requirements of equipment, and whether vibration or noise are likely to present problems.

Generally, formal design is required if a scheme includes:-

- Any element of centralisation with plant or equipment serving more than one dwelling, such as a group of sheltered flats.

- Natural draft, open flue appliances, especially if served by factory insulated chimneys or where any form of mechanical extract ventilation is proposed.

- Complex fire detection and warning systems (i.e. a Grade A, B or C system to BS 5839-6: 1995).

- Whole house heat recovery or other ducted ventilation systems.

- A lift installation.

Well planned, carefully laid out services

COMMON MISTAKES

- No design brief for services prepared or provided.

- Services not considered early enough in design process.

- Over reliance on 'specialist' design by sub-contractors or product suppliers.

- Failure to appreciate when 'rule of thumb' should give way to formal design.

Inadequate Design of Open Flues

THE PROBLEM

RATING **2 – 4**

Natural draught gas, oil and solid fuel burning appliances draw the air they need for burning from the space in which they are located, expelling the products of combustion to the outside air via an 'open' flue (i.e. without the assistance of a fan). Open flues work because as hot gases rise, cold ones must take their place, causing an upward movement of air (a 'stack' effect). This means that hot gases must be able to pass freely along the length of a flue and escape without cooling. If allowed to cool, they will lose buoyancy and 'spill' back into the room (the gases are highly toxic); if they cannot escape, the flue will not 'draw'. The performance of a flue is dependent on its materials, size and configuration, and the position of its outlet; designers must account for all of these.

GUIDANCE

Materials: Flues can be constructed of masonry or metal, though the greater the insulation the better the performance. Masonry flues – which should be lined or constructed of refractory concrete – have a greater thermal capacity and remain warm longer, aiding restart. However, metal flues become hotter, increasing the draw.

Size: Flues should be of the following minimum sizes (and at least as big as the outlet of the appliance it serves):–

- **Gas:** 12,000mm^2 (round flue) or 16,500mm^2 (rectangular flue with a minimum dimension of 90mm) for gas fires (other than fuel effect type). Fuel effect fires should be treated as solid fuel appliances, unless otherwise stated by the manufacturer.

- **Oil:** 100, 125 or 150mm diameter (or equivalent area), depending on the rated output of the appliance (i.e. whether below 20kW, 32kW or 45kW).

- **Solid fuel:** 125, 150 or 175mm diameter (or equivalent area), depending on the rated output of the appliance (i.e. whether below 20kW, 30kW or 45kW) for 'closed' appliances (150mm if coal-burning). For 'open' fires, 200mm diameter if the fireplace recess is less than 550mm x 500mm, otherwise 15% of the area of the recess. Increase by 25mm if flue is offset.

Configuration: Vertical wherever possible. Bends should not make an angle of more than 45° with the vertical (30° for solid fuel). There should be at least 600mm of vertical flue above any draught diverter. Horizontal runs are unacceptable, other than for back-outlet solid fuel appliances (maximum distance 150mm).

Outlets: Should be located at least 100mm above roof level (600mm if within 600mm of the ridge of a roof with a pitch of more than 10°) and fitted with a terminal.

A supply (free area) of air is essential. For gas appliances with a rated *net input* up to 70kW, 500mm^2 for each kW of input over 7kW; for oil and 'closed' solid fuel appliances with a rated *output* up to 45kW, 550mm^2 for each kW of output over 5kW; for open fireplaces, 50% of the throat area. *The supply may be seriously reduced by extract fans anywhere within the building.*

An (unacceptable) horizontal flue

COMMON MISTAKES

- Single wall (i.e. uninsulated metal) flues used externally.

- Flexible flues unsupported in roof voids.

- Multiple bends in flues.

- More than 150mm length of horizontal flue to back outlet solid fuel appliances.

- Insufficient air supply.

- Failure to consider effect of extract fan, not just within rooms containing appliances, but within whole building.

REFERENCES

BS 5410–1:1997 cls.9 & 10

BS 5440–1:1990 cls.3 & 4

BS 5440–2:2000 cls.4–6

BS 5871–1:1991 cls.8 & 9

BS 5871–2:1991 cls.8 & 9

BS 5871–3:1991 cls.7 & 8

BRE IP 7/94

BG DM3

CPDA 1996

Plant or Components Unsuited to Water Supply

The movement and delivery of water is one of the fundamental reasons for the existence of building services. Nourishment, hygiene and warmth are amongst the most basic of human needs and all depend on water. It is essential that those responsible for the design of services installations match this demand for water with the available supply. This means ensuring that plant and components are compatible with the physical and chemical properties of the water and its pattern of usage, if problems such as over (or under) pressurisation, premature scaling or bimetallic corrosion are to be avoided.

GUIDANCE

Plant and components should be selected on the basis of the properties of the water and how it is to be used. The following information should be established:-

- The quantity of water available, its supply pressure, and its required pressure and temperature in service.
- The pH value of the water (degree of acidity or alkalinity) and its 'hardness' (calcium and – to an extent – magnesium content).
- The likely frequency and maximum water demand for both supply and heating systems.

Water softening (chemical *not* electric) should be provided where the supply has a temporary hardness of 200ppm (parts per million calcium carbonate) or more and direct-feed equipment such as combination boilers or unvented cylinders are used (lime scale soon blocks small-bore pipes and can foul safety valves).

Combination boilers (not suitable where there is a high or simultaneous demand for hot water) must have sufficient water pressure to operate flow control switches. Larger boilers may need 22mm service pipes.

Copper hot water storage cylinders should have sacrificial anodes if water is soft and acidic (low pH) or with high ferric content, or when installed in a mixed metal system. To prevent splitting or distortion, the grade of cylinder must suit the head (pressure) of its feed (i.e. height from base of cylinder to water level in tank).

Unvented cylinders have maximum operating pressures; pressure reducing valves have to be fitted to the mains supply.

Open vented water supply systems should be designed to ensure that demand does not exceed supply (to prevent water and air being 'pulled down' the vent). In single storey buildings with pipes above ceiling level, the size of the *supply* pipe should be increased. In three storey houses (or higher) where the cylinder is on the top floor and there is high water demand on the ground floor, the size of the *vent* should be increased.

Copper pipework should not be connected to iron or steel (mild or galvanized) fittings due – depending on flow direction and use of inhibitors – to the risk of bimetallic corrosion.

A corroded copper ball valve

COMMON MISTAKES

- No water softening fitted where direct feed appliances used in hard water areas.
- Grade of hot water storage cylinder not matched to head (pressure) of water.
- Cylinders not fitted with sacrificial anodes (if required).
- No check valves to mains supplies feeding unvented hot water systems.
- Copper connected to iron or steel fittings.

REFERENCES

BS 5449:1990 cls.8–26

BS 6700:1997

CDA TN 39

IOP PESDG Sections A & B

Badly Planned Ventilation Systems

THE PROBLEM

RATING 2 – 3

Increased expectations of comfort and energy efficiency have increasingly led to the adoption of construction techniques that aim to minimise air leakage (e.g. draught stripping). However, this 'sealed' approach to building has come about at the same time as a decline in the use of open flues, the result being that the removal of airborne moisture and the maintenance of internal air quality cannot be achieved by natural means alone. Supplementary ventilation – mechanical or passive – is a necessity, though only effective if planned at the outset. The siting of extract points and inlet or outlet terminals, duct layout and the supply of outdoor air must all be considered if damp and foul air are to be avoided.

GUIDANCE

Mechanical ventilation can be intermittently or continuously operated. The former involves little more than individual extract fans located in kitchens and bathrooms discharging to the outside air, directly or ducted to an outlet. Continuous systems utilise a single, slow-running fan ducted to serve all extract points and discharging to a single terminal. Such systems can be balanced by mechanical *supply* ventilation, warmed – via a heat exchanger – by the heat of the exhausted air (known as 'whole house' systems). Passive stack ventilation (PSV) systems exploit the buoyancy of warm air rising through vertical ducts to provide a fixed level of ventilation without the need for a fan.

For all types of system, extract points should be sited close to sources of moisture, but as far away from air inlet points and as high as possible. Terminals must avoid flue or soil stack outlets.

In mechanical systems:–

- Ductwork should be as straight as possible, with fans sized to allow for all junctions and bends (the cross sectional area of ducts may need to be increased). Horizontal ducts should slope away from fans. Flexible ducts must be supported.

- For whole house systems, there must be at least 2m separation between the exhaust and the intake, except where specially designed combined terminals are used. On windy sites, both terminals should be on the same side of the building.

In PSV systems:–

- Ducts should be vertical with no offsets over 45°, with no interconnection between rooms (i.e. one room, one duct). Bends should be swept, not angled. Rigid ducts are preferable.

- Terminals should be within 1.5m of ridges, at or above their highest point. Roof slope vents must not be used.

A separate supply of outdoor air is required for all except whole house systems. Controllable background vents (e.g. trickle vents: 8000mm² in habitable rooms, 4000mm² elsewhere) will usually suffice, with 5–8mm gaps below doors or low-level transfer grilles serving internal rooms or rooms without windows.

Poorly planned ductwork

COMMON MISTAKES

- Extract points too close to incoming air supply.

- Flexible ducts unsupported or constricted by fixings.

- No account taken of duct resistance when sizing fans.

- Offsets in PSV ducts more than 45° from vertical.

- PSV outlet terminals further than 1.5 metres from or lower than ridge.

- No background ventilation to kitchens or bathrooms.

- Insufficient outdoor air supply to internal bathrooms etc.

REFERENCES

BS 5720:1979 cl.3
BS 5925:1991
BRE Digests 170, 306 & 398
BRE IP 18/88, 21/89, 13/94 & 6/95
BRE Report 162

Inadequate Design of Waste or Rainwater Disposal Systems

THE PROBLEM   RATING 2

Although the collection and disposal of waste and rainwater is — in essence — simply a matter of allowing water to run through an arrangement of pipes under the influence of gravity, failure to consider some of the basic rules of hydraulic design often results in installations that suffer problems in service. Sometimes systems (or parts of systems) are too small for the amount of water or discharge they have to carry, causing siphonage or — in the case of gutters — surcharge. At other times, their configuration or layout causes fluctuations in air pressure and the loss of trap seals or the formation of airlocks. Designers must therefore ensure that pipes and gutters are of sufficient size, runs of pipework and guttering are not too long and laid to falls, downpipes and soil stacks are correctly positioned, and that due account is taken of points where hydraulic loads could become concentrated, such as at valleys or multiple waste water-connections.

GUIDANCE

Waste pipework should always connect to a discharge ('soil') stack, except at ground level where discharge directly to the drainage system (or direct via a gulley if the pipe is not connected to a wc) is acceptable. *Unventilated* branch pipes serving wcs should be at least 100mm in diameter, no longer than 6m and laid to a fall of at least 9mm per metre. Sinks or baths may be served by 30mm or 40mm pipes sloping at between 18mm and 90mm per metre and not exceeding 3m or 4m in length respectively. The gradient of a pipe serving a washbasin depends on its length, which must not exceed 1.7m for a 32mm pipe, 3m for one of 40mm.

Soil stacks should be vertical and with a diameter of at least 75mm, with a bend of at least 200mm radius at its base. No branches should be within 450mm of the invert (bottom) of the stack, and branches on opposite sides of the stack should be offset (110mm if the stack is 100mm in diameter, 250mm if 150mm diameter); branches opposite wc connections should enter the stack via a parallel or angled connection at least 200mm lower. Offsets in the 'wet' part of the stack are only acceptable (but still not desirable) in low-rise buildings, and where there are no branch connections within 750mm. All soil stacks should be ventilated to the open air unless an automatic air admittance valve (aav) backed by a BBA certificate is used.

Stub stacks are short unventilated stacks that can be used to connect groups of ground floor appliances to the drains, as long as no wc is connected higher than 1.5m above the invert level of the drain, and no other branch is higher than 2.0m.

Rainwater systems for areas of roof up to about 40m²: 100mm half round gutters with 63mm outlets are suitable. Gutters and downpipes for larger roofs should be sized by calculation, with allowances made for increased flows at roof valleys, below large areas of walling (especially glazing) and other features that will concentrate run-off. All gutters should fall towards outlets, unless specifically designed to run 'deep flow'.

Excessively long branch pipes

COMMON MISTAKES

- Long runs of waste pipework without ventilation.

- Tails of traps not extended where connected to waste pipes of greater diameter.

- External quality rainwater downpipes or soil stacks used internally (joints not sealed).

- Concentration of rainwater at valleys or run-off from vertical surfaces (especially glass) not accounted for in sizing of gutters.

REFERENCES

BS 5572:1994 cls.6 & 7

BS 6367:1983 cls.5–9

BRE DAS 55 & 143

BRE GBG 38

IOP PESDG Section C

Insufficient Support to Pipes or Gutters

THE PROBLEM   RATING 2

Pipework and guttering in service is susceptible to sagging, distortion and buckling due to the physical forces arising from the weight or pressure of their contents, thermal movement, vibration and external impact. Water can be prevented from flowing, joints may fail due to their being subjected to stresses for which they are not designed, and whole sections become dislocated. It all depends on how well the pipes or gutters are supported: the right type of support at the correct centres must be specified if liquids (and sometimes solids) are to be circulated or discharged without leakage or overflowing.

GUIDANCE

Supports should typically be spaced as follows:–

Pipe type	Vertical spacing	Horizontal spacing
Copper water pipe 15mm	1.8 metres	1.2 metres
Copper water pipe 22mm	2.4 metres	1.8 metres
Plastic water pipe	as manufacturer's recommendations	
Plastic waste pipe 32–40mm	1.2 metres	0.5 metres
Soil stacks/branches 75–100mm	1.8–2.0 metres	0.9 metres
Gutters & downpipes (all types)	1.8–2.0 metres	0.9–1.0 metres

Additionally:–

- Extended supports will be required where pipes are to be insulated (to prevent the insulation being compressed between the pipe and the surface to which it is fixed).
- Rigid fixings, especially at tee branches, should be avoided.
- To protect connections, some allowance should be made for the movement of appliances relative to pipework (e.g. vibration of washing machines).
- Soil stacks and rainwater downpipes should be supported directly below each connection and coupling.
- The bases of all soil stacks should be supported (i.e. where the slow bend rests within the ground).
- Extra support should be provided to gutters within 150mm either side of corners.
- Movement joints may be required to long runs of plastic pipe or guttering (can expand up to 5mm per metre when warmed by the sun).

Support to the corner of a plastic gutter

COMMON MISTAKES

- Supports too far apart.
- No extra support provided to corners of gutters.
- Soil stacks not supported at connections or at base.
- Pipes or gutters fixed to an uneven background (causes clips or brackets to 'spring').
- No provision for thermal expansion of gutters, downpipes or stacks.
- Too little space for insulation behind pipes.

REFERENCES

BS 5449:1996 cl.31
BS 5572:1978 cl.9.2
BS 6700:1997
BRE DAS 41, 42 & 101

Walls, Floors or Roofs not Designed to Support Services

THE PROBLEM   RATING 1

Services, appliances and fittings can be very heavy, especially if they are made of or contain cast metal components (e.g. boilers), hold water (e.g. radiators. storage tanks and cylinders) or are required to support the weight of people (e.g. wc pans and baths). Robust fixings and firm support are essential if they are to remain stable and not suffer or cause damage, which is generally not a problem where walls are of dense masonry and floors of solid construction. However, it can be a problem where walls are framed, dry-lined or constructed from aerated concrete blocks, or where floors are overlain with a 'floating' layer. Vibrations set up by appliances (e.g. washing machines and dishwashers) can also present problems, especially for timber floors and decks. It is therefore important that designers identify the locations of all heavy appliances and fittings at the outset and explicitly include provision for secure fixing and support within their designs.

GUIDANCE

Framed walls: Additional studs and noggings to support fittings should be provided, although their positions will need to be carefully recorded and marked on linings if final fixings are to be accurately placed. It may be better – and more flexible – if a 12 or 18mm layer of plywood is fixed behind the lining (plasterboard or render on metal lath), which will also spread the load of the fitting better. Less heavy fittings may be fixed to a double layer of plasterboard (with special fixings), though this is not advisable if any 'pulling' forces are expected.

Dry lined walls: All fixings should penetrate through the lining and into solid masonry. Wall mounted boilers will need to be fixed on spacing brackets (templates are usually provided for installation).

Aerated concrete blockwork: Special fixings are required where heavy fittings are to be mounted on aircrete block walls. Individual manufacturers should be consulted.

Timber floors: Baths on timber floors should be set on cross bearers on top of the floor deck (to spread the load across joists); to prevent the failure of sealant joints, rims should be supported by wall battens. Floor standing boilers should be placed on rigid decking with a smooth, heat resistant and waterproof surface. Where vertical stability is essential, such as with cylinders and tanks, cross bearers should be provided. Where appliances or fittings sit on floating layers, additional support battens should be provided at the edges (see also p.64).

Roofs: Those responsible for the design of the structure should be advised of any abnormal loads on ceiling joists. Water tanks should always sit on purpose made stands that spread the load across ceiling joists or trussed rafters (see also p.38).

Appliances: Washing machines and dishwashers require rigid support and space to allow for vibration, particularly if installed in a cupboard where services in close proximity might be damaged.

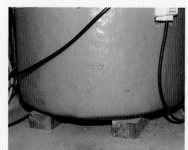

Support to a hot water cylinder

COMMON MISTAKES

- No account taken of fixing into aerated concrete blocks.

- No support to rims of baths.

- Timber joists not designed to take large storage cylinders, floor standing boilers or storage heaters.

- No additional support to washing machines, wc pans etc. on floating floors.

- Tanks in roof spaces not supported by tank stands, properly constructed with a solid board under the tank.

- Washing machines on plinths not bolted down or provided with edge restraint.

REFERENCES

BS 5268–3:1997 cl.7.5

Failure to Take Account of Operating Environment

THE PROBLEM   RATING 2

Many types of services installation are sensitive to the physical environment in which they are located. Temperature, humidity, direct sunlight and airborne particles can all — under certain circumstances — adversely affect their operation and cause damage to systems or components. Frozen pipework within roofspaces, rusting metal, plastics that have turned brittle, and the premature failure of motor units clogged with dust are all examples of such damage. The design of services installation and the selection of components must account for the environmental conditions under which they are expected to operate.

GUIDANCE

Water supply systems: Pipes and storage tanks in unheated areas should be insulated to prevent freezing or condensation. Properly fitted jackets that can be easily removed and refitted for maintenance and servicing should be used for tanks, storage cylinders etc. Care should be taken to ensure that pipes carrying drinking water are not over-warmed (i.e to over 25° centigrade) due to the risk of bacterial growth, and water in storage must not be warmed to between 20° and 50° centigrade due to the risk of legionella. Beware of siting tanks in 'warm' roofspaces that also contain uninsulated heating pipes.

Plastic pipework: Polythene pipe becomes brittle when exposed to UV light. Exterior pipework should be uPVC. Waste pipes passing through unheated spaces such as cold roof voids may require trace heating to prevent freezing.

Ventilation: Ductwork passing through unheated spaces such as roof voids must be insulated. Cold air ducts should also be wrapped in a vapour barrier. In certain areas (e.g. kitchens where fat-frying is regularly carried out), extracts may need to be fitted with filters to prevent fan units becoming clogged with dust or grease. Automatic air admittance valves used to vent plumbing systems should also be insulated and protected from dust.

Lift installations: Motor rooms serving hydraulic lifts may require low temperature space heating to maintain the viscosity of hydraulic fluids during cold weather. The pipes carrying the fluid may also need to be insulated or trace heated if the motor room is remote from the lift it serves. All types of lift motor room should be ventilated to prevent the build up of condensation (risk of electrical short circuits and corrosion of metalwork).

Electrical equipment: Cables laid under insulation must be sized to prevent overheating, and expanded polystyrene insulation should not be used next to PVC sheathed electrical cable (the polystyrene and the PVC can react causing loss of plasticiser, indicated by a sticky blue/green deposit). Recessed light fittings – especially in low voltage systems – should be ventilated, as should all transformers. Standby batteries should be protected from extremes of temperature. External light fittings must be sealed.

Corrosion due to condensation

COMMON MISTAKES

- Pipes and tanks in roof spaces not insulated.
- Cold water supply pipes run in uninsulated ducts (risk of condensation).
- Ventilation ducts within roofspaces not insulated.
- No heating to hydraulic lift motor rooms.
- No ventilation provided to lift motor rooms.
- Heat sensitive electrical equipment sited next to uninsulated heating or hot water pipes.
- Cables within depth of insulation not down-rated.

REFERENCES

BS 6700:1997 cl.2.6
BS 7671:1992
BRE DAS 62, 101 & 109
BRE Digest 398
BRE Report 262 p.29

Lack of System Protection

Building services are — by their very nature — dynamic in operation. They move (or cause movement), cause variations in temperature, generate electric fields and produce controlled emissions or spillage. All of these phenomena are a potentially direct threat to health and safety, and each has the capacity to cause damage to both the fabric of a building and the services installations themselves. Hence the necessity for designers to consider the whole array of safety valves, trip switches, temperature regulators, overflows and other protective devices that form a significant part of any services installation that might threaten the physical well-being of persons or property.

GUIDANCE

Protection of persons: This is the main reason for much of the system protection included within services installations. Specific points to note within a domestic context include:–

- Heat shields should be provided where people could come into contact with hot gases or surfaces (e.g. where balanced flues are within 2 metres of the ground). It may also be necessary to shield or insulate standard heating and hot water pipes in buildings used by the frail elderly or other persons who are particularly vulnerable to burns or scalding.

- Tanks should have properly fitting lids and shielded vents to ensure that water supplies are not contaminated.

- Electrical systems must be fully earth bonded with suitably rated circuit breakers fitted throughout (i.e. miniature or residual current circuit breakers).

- Carbon monoxide detectors should be considered for all locations where open flue heating appliances are installed.

Protection of building fabric: Separation gaps will be required between metal flues and any combustible building fabric. Shielding must also be provided where hot gases could come into contact with plastics or other heat-sensitive material. Overflow and warning pipes – which should be readily visible and able to cope with a full supply flow – should be fitted to all storage tanks, cisterns, and sanitary fittings; also to parapet or valley gutters. The provision of drip trays should be considered if disruption due to leaking tanks may be unacceptable (e.g. care homes).

Protection of services installation: A well designed installation will include 'leakage' points that ensure plant or components are not damaged if any part of the system fails. Examples include:–

- Hot water vents as found on heating, hot water systems or boilers (vents should discharge into a tank or drain to prevent splashing and scalding).

- Release valves fitted to pressurised hot water systems.

- Shut off valves and isolation switches, all of which must be readily accessible in the event of an emergency.

Earth bonding to a metal bath

COMMON MISTAKES

- Unprotected balanced flues close to plastic plumbing.

- Cold water warning and overflow pipes not positioned to discharge away from building, nor protected from freezing.

- Spillage detectors not fitted to open flue gas appliances.

- No protection to external electrical supplies.

- Inappropriate use of RCCBs (e.g. to cookers, freezers and other fixed appliances that might be 'wet' in service).

REFERENCES

BS 7430:1991

BS 7671:1992

BRE DAS 61, 62 & 92

BRE GBG 31

IOP PESDG Section M

No Formal Provision for Commissioning and Testing

THE PROBLEM

RATING **3 – 4**

Services installations are designed to fulfil specific functions, just like any other part of a building. However, whereas it is almost always obvious at the outset whether a wall or other 'static' element is fulfilling its designed function, the successful performance of nearly all services installations only becomes apparent after time. For instance, the success or failure of a heating system completed in summer may only become apparent during the cold of winter, which – from a user's point of view – is neither practical or desirable. Hence the need for all services to be formally commissioned ('set up') and tested before they are brought into operation. But commissioning and testing require time and resources; they cannot be simply tacked-on to a project at the last minute. Designers must consider commissioning and testing as an integral – and formal – part of their work if damage as a result of defects within a services installation is to be prevented.

GUIDANCE

Specifications for commissioning and testing must form a part of the contract documentation for all projects, and – ideally – should be implemented during the construction programme, not after all works have been completed.

The following items might be included in respect of the commissioning of domestic services:–

- Flushing and sterilisation of water supply and heating pipes.
- Setting of water levels within cisterns and storage tanks.
- Adjusting the flow at taps and other water draw-off points.
- Balancing boilers and heating systems.
- Setting timers, thermostats and humidity-sensitive controls.
- Setting the speed and overrun times of extract fans.

Things that might be tested or checked include:–

- The effectiveness of flues serving natural draft appliances.
- Supply volume and heat recovery time of hot water.
- Hot water temperature at the point of delivery.
- Output temperature of all radiators.
- Functioning of pressure and temperature relief valves.
- Air tightness of ventilation systems.
- Performance of waste pipework and traps under pressure.
- Depth of trap seals maintained after full or multiple discharge.
- Rainwater gutters and pipes for leakage.
- Continuity of electrical systems and earth bonding.
- The polarity and loading of electrical circuits.
- Activation of smoke or heat detectors and fire sounders.
- Life of battery packs to emergency escape lighting.
- Performance of overflows and warning pipes.

Balancing the temperatures on a boiler

COMMON MISTAKES

- Failure to flush pipework (to remove flux and debris) before testing.
- Masonry flues not tested for blockages or leaks.
- Air tightness of ventilation systems not tested.
- Temperatures of flow and return to heating systems not balanced on a radiator by radiator basis.

REFERENCES

BS 5449:1995 cls.38–41
BS 5871–1:1991 cl.15
BS 5871–2:1991 cl.14
BS 5871–3:1991 cl.13
BS 6700:1997 cls.3.1.10–3.1.12
BS 6798:1987 cl.14
BS 7671:1992
BSRIA AH 87/1, 89/2 & 92/3
CDA TN 33

Below Ground Drainage and External Works

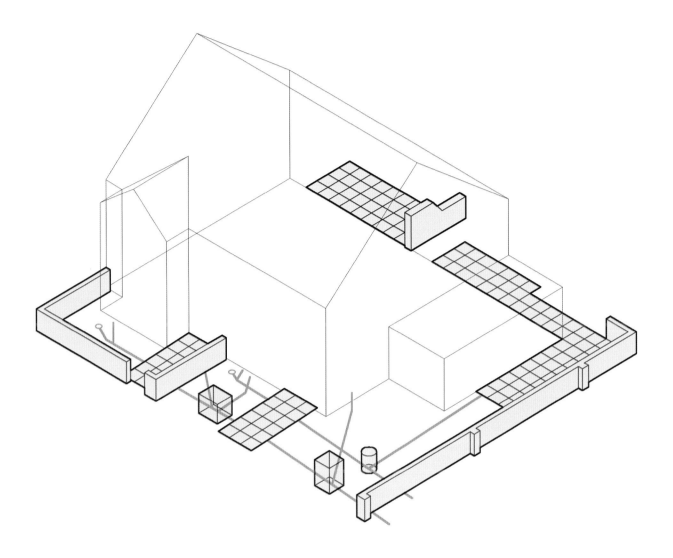

Defects and their Consequences

DEFECT	CONSEQUENCES			
	A	B	C	D
LACK OF ADEQUATE SITE INVESTIGATION	3	3	3	3
EFFLUENT RUN TO UNSUITABLE GROUND	3			
INCORRECTLY LOCATED OR SIZED RAINWATER SOAKAWAYS	4			
NO PRECAUTIONS TO ENSURE STABILITY OF DRAINS IN POOR GROUND	3	4	3	
POORLY PLANNED FOUL DRAINAGE	3			
LACK OF PROTECTION TO SHALLOW DRAINS		3		
POOR SPECIFICATION AND DETAILING OF MASONRY WALLING			3	3
INADEQUATE DRAINAGE TO REAR OF RETAINING WALLS			3	
FAILURE TO ADEQUATELY CONSIDER FOUNDATIONS TO PAVEMENTS		3	3	4
INAPPROPRIATE OR POORLY BEDDED SURFACING TO PAVEMENTS			3	4

The four columns labelled A to D refer to the four types of consequence detailed on the opposite page.

KEY TO HAPM RATING

1 Low probability of defect occurring, and only likely to have minor consequences.

2 Low probability of defect occurring, though with potentially serious consequences.
   or
   Reasonable probability of defect occurring, though only likely to have minor consequences.

3 Reasonable probability of defect occurring and with potentially serious consequences.
   or
   High probability of defect occurring, though only likely to have minor consequences.

4 High probability of defect occurring and with potentially serious consequences.

Consequences

A

SURCHARGE OF DRAINS AND FLOODING

The first visible sign of defects in the design of a drainage system is often the 'smell of drains' as water backs-up within blocked pipelines and inspection chambers, or begins to leak into the surrounding ground. Evidence of failure may also appear as damp patches forming in the vicinity of inspection chambers and the bases of discharge stacks, or the 'ponding' of surface water around gullies. More extreme problems may result in the displacement of inspection covers or the failure of the seals to traps serving appliances, and even wholesale flooding with rainwater or foul effluent; the latter may well present a threat to health and render properties uninhabitable, even those not visibly affected.

B

FRACTURE AND DISPLACEMENT OF DRAINS

Although often not discovered until the occurrence of surcharge and flooding, the fracture and displacement of pipelines is generally an indication that there are defects in the design of the below ground drainage system. Rigid pipes subject to excessive loads may be cracked or broken and plastic pipes deformed in section. Pipes subject to movement may sag and lose their gradient, developing 'back' falls that prevent the discharge of liquids and solids; movement can also result in the dislocation of joints and couplings, leaving the contents of the drain — raw sewage or surface water — free to run to ground.

C

DISPLACEMENT OR CRACKING OF WALLS OR PAVEMENTS

Structural movement can cause walls to lean or bulge, or the formation of dips and hollows in paved surfaces, often preventing water from draining and leaving the edges of inspection covers etc. standing proud. Such movement can also cause cracking and opening-up of mortar joints and the junctions between elements, or the appearance of potholes and fissures. Thermal movement may also cause cracks, normally running through — rather than around — individual elements.

D

PREMATURE DETERIORATION OF EXTERNAL WORKS

External works frequently have to endure what — in building terms — are extremely aggressive service conditions. Wetting, freezing, poor ground and mechanical wear and tear may all take their toll, causing the premature decay of materials or components. Surfaces may erode, soften and become brittle or detached (spall). Materials such as mortar and concrete may lose their inherent strength, necessitating the extensive rebuilding of masonry walls; likewise the materials and systems used to make-up foundations to pavements, the repair of which can be costly and disruptive.

Lack of Adequate Site Investigation

THE PROBLEM   RATING **3**

It is sometimes forgotten that drains and external works must be matched to ground conditions in exactly the same way as foundations. However, site investigations are often deficient in this respect, and the importance of issues such as weak, loose or unstable soil, and the level and behaviour of ground water underestimated, especially as regards the stability of vehicular or pedestrian surfaces. Nor is it sometimes appreciated that the information gleaned from trial pits, bore holes, etc. is — in itself — not always enough; the site must be seen as a whole if the dislocation, fracture and — in extreme cases — collapse of drains, pavements, freestanding and earth-retaining walls is to be avoided.

GUIDANCE

The objectives of the site investigation – insofar as it may relate to the design of below ground drainage and external works – must be decided at the outset. The investigation should progress through the following stages:–

Excavating a trial pit

- **Desk study:** At the very least this should entail a study of old maps and geological information, local records, and any previous site investigations. Meteorological data and information held by the environment or river authorities may also be of value. Particular attention should be paid to old ditches and watercourses, rubbish pits and any other features that suggests unstable ground or waterlogging.

- **Site reconnaissance:** This involves a visual assessment of the site and its locality (i.e. a 'walk over' survey). Desk study information is checked and landscape features that may influence ground conditions (e.g. rivers) are recorded. Nearby cuttings or quarries can help in assessing the likely ground conditions; flood defences and the siting of older properties might indicate a waterlogging problem.

- **Detailed examination:** It is at this stage that the ground investigation is carried out. The positioning of trial pits and bore holes should enable the nature of any soil in which drains are to be laid or which is to support roads or paving to be fully described. Ground water levels, California Bearing Ratios and – if appropriate – percolation values must all be established; on-going monitoring may be required.

- **Follow up investigations:** The findings of the ground investigation should be reviewed as and when the ground is disturbed by construction operations.

Although stages may overlap, the importance of the desk study and site reconnaissance being carried out prior to the ground investigation cannot be stressed too highly. The failure to carry out such an exercise makes it very difficult to ensure that the nature of any ground investigation is proportional to the scale of the project, and that design decisions are not based on unrepresentative or misleading information.

COMMON MISTAKES

- Insufficient investigation of ground when conditions for support of buildings already established as poor.

- No desk study.

- Failure to grasp distinction between site investigation and ground investigation.

- Levels, behaviour and annual variations of ground water not established.

- No follow–up or review during construction.

REFERENCES

BS 5930:1981

BS 6297:1983 cls.4 & 15.3.2

BS 7533:1992 cl.4.1

BS EN 752–3:1996 cl.8.4

HA DMRB7:HD 25/94 cls.2 & 4

Effluent run to Unsuitable Ground

THE PROBLEM   RATING **3**

A lack of infrastructure in some rural areas means that it is not always possible for foul drainage to be run to a public sewer. Sewage must either be stored in a cesspool (a covered watertight tank) until it can be removed for treatment elsewhere, or it must be treated on site. Treatment can take a number of forms, and can result in effluent (the liquid resulting from the anaerobic breakdown of raw sewage) that is pure enough to discharge – under licence – into a river or tidal waters. However, it is more common on small developments for sewage to be allowed to settle and separate in a septic tank, and the resulting effluent disposed of via a percolation area (a system of field drains designed to distribute the effluent over a large area) or a soakaway pit. This means septic tanks are only appropriate if a suitable area of porous ground with a low water table is available. Otherwise, the effluent will not drain away, filling the septic tank and backing-up along the drains.

GUIDANCE

Before any decisions can be taken on the disposal of effluent to the ground, the level of the water table *in winter* and the percolation value of the soil (the average time – in seconds – it takes water at the level of the proposed land drains to drop by 1mm) must be established. Investigations must be in accordance with the methods set out in BS 5930 and BS 6297 respectively, with particular note being paid to time scales and weather (determination of the water table from a simple pit requires it to be left open for a period of time, and no percolation test should be carried out during periods of heavy rain, severe frost or drought). Account should be taken of seasonal variations in rainfall or the levels of any nearby rivers.

A percolation area (sub-surface irrigation) should:–

- Only be used if the percolation value of the soil is less than 100 seconds, although soil with a value up to 140 seconds may be suitable if 'underdrains' are provided (a second set of drains set below the first and designed to convey surplus drainage to a ditch or watercourse).

- Not be used where the water table in winter rises to within 1000mm of the *invert* level of the field drains.

- Have a trench area of 0.25 x the percolation value x the number of persons served by the septic tank (trenches should be 300mm to 900mm wide with at least 2m of undisturbed ground between parallel trenches, and deep enough to ensure the drains are at least 500mm below ground level).

Soakaway pits ('soakpits') – possibly no more than an excavation filled with brickbats from which the effluent may percolate into the surrounding ground – are only suitable for porous soil (e.g. gravel or chalk) and where the effluent can be discharged *above* the winter water table. However, their use is to be discouraged due to the fact that discharging effluent direct to highly porous ground can result in pollution of the water supply (final treatment actually relies on filtration by the ground).

Installation of private treatment plant

COMMON MISTAKES

- Proposals not discussed with relevant authority (e.g. the Environment Agency).

- No account taken of level of water table in winter.

- Percolation test carried out during abnormal weather

- No account taken of effect of rainfall or nearby rivers, watercourses, etc. on water table and percolation value.

- Percolation test not carried out in exact location of proposed drains

REFERENCES

BS 6297:1983 cls.4, 6.2.2.1 & 15.3

NSAI SR6:1991

CIRIA TN 146

CIRIA SP 144/L2

Incorrectly Located or Sized Rainwater Soakaways

THE PROBLEM   RATING 4

Soakaways are often used to assist in the disposal of rainwater from buildings. They enable the construction of new developments that are remote from a public sewer, where existing sewers do not have the capacity to receive further water, or where ground water levels must be maintained. But the flow of rainwater to a soakaway is not something that occurs at a steady, continuous rate. Storms come and go with varying degrees of rapidity and intensity, and hence a soakaway must be able to store water as well as enable its dispersal; location and size must be carefully considered in relation to site conditions if surcharge and flooding are to be prevented.

GUIDANCE

A soakaway is simply a pit in which water can collect before percolating into the adjacent ground. It can be square or circular in plan, or it can take the form of a trench. Small pits (i.e. draining areas no larger than 100m²) may be no more than a hole filled with rubble (to support the sides of the excavation) or they may be unfilled and with a perforated lining surrounded by granular material; large pits must always be lined.

Porous soils (e.g. chalk, gravel or fissured rock) are most suited to the use of soakaways. However, there is nothing to prevent their use in any type of soil, so long as the site is large enough to accommodate the required volume of soakaway:-

- Above the level of the water table and – on sloping ground – sited to avoid waterlogging lower down.

- No closer than 5m to any building, nor in a position where the ground beneath foundations could be adversely affected.

- As far away as possible from any septic tank and not so that stormwater will run towards any percolation area.

The effective volume of a soakaway – which excludes the space occupied by any rubble – is measured between its base and the invert level of the inflow pipe, and must be no less than the difference between the volume of water that flows *in* and that which flows *out* (infiltrates) into the soil while it is raining.

Inflow is dependent on the area drained and the amount of rain falling during a storm. Outflow is determined by the internal area of the sides of the soakaway (the base is presumed to be ineffective in dispersing water), storm duration and how easily water can infiltrate the soil. Calculation methods are given in BRE Digest 365, with details of how to determine the soil infiltration rate by measuring the time it takes water to drain from a trial pit.

It is important to note that the procedure for designing a soakaway starts with assumptions as to its volume and depth, assumptions that must be reviewed following on-site tests. A soakaway may not work if it cannot discharge at least half its volume within 24 hours even if (in theory) it is correctly sized.

A trial pit to determine the water table

COMMON MISTAKES

- No investigation of level and fluctuations of water table.

- Soakaway volume and depth measured from ground level rather than level of inflow.

- Siting of soakaway likely to cause waterlogging down slope, or to adversely affect ground under foundations.

- Volume of rubble fill not omitted when calculating soakaway size.

- Trial pits for assessment of soil infiltration rate not taken down to proposed depth of soakaway.

- No assessment of time taken for soakaway to drain 50%.

REFERENCES

BS 8301:1985 cl.8.4

BS EN 752–3:1996 cl.8.4.7

BRE Digest 365

BRE GBG 38

No Precautions to Ensure Stability of Drains in Poor Ground

THE PROBLEM   RATING **3 – 4**

A drain must maintain its designed fall and remain watertight if it is to function as intended, which means that the ground in which it is laid must provide stable and consistent support. However, it is not always possible to avoid laying drains in soft, variable or inherently unstable soils; and hence the need to design pipelines and inspection chambers to accommodate movement (including differential movement) or for the soil to be 'improved'. Failure to account for poor ground can lead to the displacement, fracture or wholesale failure of the drainage system, possibly leading to a further weakening of the soil and even the disturbance of pavements or foundations.

GUIDANCE

Ideally, pockets of soft ground and 'hard' spots (e.g. rocks) should be removed and replaced with well-compacted material so as to provide a uniform foundation ('formation') for the drain. If this is not practical then:-

- Rigid pipes with flexible joints or flexible pipework laid to the maximum possible gradient should be used (steeper gradients lessen the risk of ground movement leading to 'back-falls').

- Flexible joints should be provided within 150mm of any point where a pipe passes through a wall or connects to an inspection chamber, with the first length of 'rocker' pipe being no more than 600mm (alternatively, an opening that leaves at least 50mm clearance around the pipe can be formed).

- Allowance should be made for blinding or stabilising (immediately after excavation) the formation levels of trenches in wet, fine grain soils that have the capacity to support the pipe bedding material (e.g. soft clays).

- Trenches in ground where the formation level will not support the bedding should be over-excavated by at least 600mm and a firm base formed from compacted layers of gravel or hardcore (max. size 75mm) blinded with 50mm of lean-mix concrete. Drains should be laid on at least 150mm of single size or graded granular material (200mm for socketed pipes).

Excavations in unstable ground may necessitate wide trenches and hence a greater loading on the drain; the thickness of the bedding and surround may have to be increased, especially for flexible pipes (the surround limits deflection and prevents buckling).

Where drains are laid in waterlogged ground the entire bedding system – including the sidefill and main backfill – should be wrapped in geotextile material (to prevent fine material being washed from the surrounding ground and causing a loss of support to the drain and its bedding).

In certain situations, the ground conditions may be so poor that pipelines will require continuous support (e.g. set on piles), in which case specialist engineering advice should be sought.

Bedding drains in waterlogged ground

COMMON MISTAKES

- Drains in unstable ground laid to minimum gradients (i.e. no provision for ground movement).

- No rocker pipes at walls or inspection chambers.

- Pipes not laid on granular material (i.e. laid direct on base of trench) or granular material not thick enough.

- Over-excavation not taken below inspection chambers, access eyes, etc.

- No membrane to prevent washing of fine material from waterlogged ground.

- Unsupported pipelines laid in exceptionally poor or variable ground.

REFERENCES

BS 8301:1985 cls.6.7, 14 & 18.2

CPDA TN 01 & TN 02

CPDA 1999 pp.11 & 12

Poorly Planned Foul Drainage

Foul drainage must be designed so as to convey and discharge liquids and solids to a public sewer, private treatment works or cesspool without leakage, blockage or surcharge. Gradient, 'line' (straightness) and the provision of access are essential factors in achieving these criteria and must be carefully considered by designers at the outset, since poor planning at this stage can result in a system which is not only hydraulically inadequate, but also difficult to maintain and service.

GUIDANCE

As much as possible of the below ground drainage system should be external, the only parts under buildings being short branches serving discharge stacks or ground floor wcs. Drains should be vented to the external air via a soil stack or a ventilating pipe near the head of the system; ventilation should also be provided near intercepting traps (e.g. at the connection to a sewer). Automatic air admittance valves should not be used in either of these situations.

Drains should be planned so that:–

* The discharge from all branches into an inspection chamber or manhole should be swept in the direction of flow of the drain, avoiding any acute changes in direction; branches should not enter at angles greater than 90°.

* Runs between junctions and access points are as straight as possible. If bends cannot be avoided (e.g. to enable the formation of a 45° junction where a branch is laid against the direction or flow) they should have long radii.

* Gradients of foul drains are not less than 1:80 for a 100mm pipe (1:40 if the anticipated peak flow is less than 1 litre per second) or 1:150 for a 150mm pipe (which must carry discharge from at least 5 wcs).

* Access points – rodding eyes, access fittings, inspection chambers or manholes – are provided at or near the head of each run of drain and at all branch junctions. Heads of runs and junctions should be no further than 12m from an access fitting (22m from a junction if the access fitting is large), 22m from an inspection chamber or rodding eye (junctions only), or 45m from a manhole; access must also be provided at any change of direction, gradient or pipe diameter.

* There is an inspection chamber or manhole within 22m of a small access fitting and 45m of a large access fitting, and that the distance between two inspection chambers or an inspection chamber and a manhole does not exceed 45m (two manholes can be up to 90m apart).

* Rodding eyes are no further than 22m from an access fitting, or 45m from an inspection chamber or manhole.

Foul drainage being laid

* Branches to manholes and inspection chambers set against direction of flow.

* Short radius bends without access for cleaning.

* Drains with low peak flows laid at gradients that are too shallow.

* No access for dealing with blockages, or clearing only possible from one direction.

REFERENCES

BS 8301:1985 cls.6, 7, 8 & 13
CPDA 1999 pp.6–9
IOP PESDG Section C

Lack of Protection to Shallow Drains

It is not always possible for drains to be laid at a depth where protection against crushing, deformation or buckling can be provided by bedding and backfill alone. Soil conditions, the practicality of forming connections (especially at the head of a run) or a desire to avoid deep excavations can all result in pipelines that are shallow enough to be easily damaged by traffic loading, construction operations (present and future), activities such as gardening and cultivation, or mechanical damage (e.g. the erection of fences). Shallow drains may also suffer damage where they pass beneath or are in close proximity to buildings. Designers must be able to anticipate when and where drains might be susceptible to damage, and be aware of what additional provisions might be necessary for their protection.

GUIDANCE

Pipes should be fully surrounded in at least 100mm of concrete (150mm or a quarter of the outside diameter – whichever is the greater – for pipes larger than 300mm) if they are:–

- Rigid and have less than 0.3m cover (0.6m if the pipe has a diameter of 150mm or greater).

- Less than 300mm below the underside of a ground-bearing floor (the encasement should be integral with the floor).

- Below the level of the foundations of a nearby building (the concrete should fill the drain trench up to the level of the underside of the footing if the foundation is within 1m).

- Within 0.9m of the finished surface of a road (the concrete should also be reinforced; alternatively, the drain may be protected with reinforced concrete 'bridging' slabs with at least 300mm bearing each side of the trench and separated from the granular surround by a compressible layer; note that the slab must be structurally capable of carrying all loads imposed on the road).

Movement joints of compressible board (e.g. 18mm thick bitumen impregnated insulating board or expanded polystyrene) should be formed in the concrete at every pipe joint except where the surround is integral with a ground slab or heavy loading dictates otherwise (however, there should be at least one movement joint every 5m). Concrete should be no weaker than grade C20 or C20P and well rammed. Trench formations must provide a firm foundation for the concrete; a blinding layer or the removal and replacement of 'soft spots' may be required.

Flexible pipes less than 0.6m below the surface of fields or gardens should be protected against damage, either by encasement or by placing over them a layer of concrete paving slabs separated from the pipe by at least 75mm of granular material.

Consideration should be given to protecting any drain that is within 1.2 metres of the wearing surface or 1.0m of the formation level of a road, or less than 0.9m below a field or garden.

A vulnerable shallow drain

COMMON MISTAKES

- Drainage planned without any reference to final ground levels.

- Flexible joints not specified for concrete surrounds.

- Lack of compressible layer between bridging slabs and surround (to prevent loads being transmitted to pipe).

- Shallow pipes below fields or gardens left unprotected.

REFERENCES

BS 8301:1985 cls.11.1.2 & 20.4
CPDA TN 04 & TN 06
CPDA 1999 pp.15–18

Poor Specification and Detailing of Masonry Walling

THE PROBLEM   RATING **3**

The exposure of freestanding and earth retaining walls to wetting from all sides and from above, means that the masonry from which they are constructed is always at risk from saturation and the leaching of salts into its pores, which gradually become clogged with microscopic crystals, leading to efflorescence and — ultimately — the exfoliation (shedding) of surfaces. Wet masonry is also susceptible to damage as the water within its pores freezes and expands at low temperatures. It is therefore essential that careful attention is paid to the quality of the materials used in the construction of freestanding or earth retaining walls, and that their detailing limits exposure to excessive levels of moisture.

GUIDANCE

Clay and calcium silicate brickwork used in the construction of freestanding or earth retaining walls should have a density of at least 1200 kg/m³ and be of no lesser quality than type MN or ML to BS 3921:1985 (clay) or class 3 to BS 187:1978 (calcium silicate) if the top of the wall is protected by an over-hang; otherwise use types FN, FL or class 4 respectively (these qualities of bricks should always be used below DPC level and for cappings; also for the entire wall in exposed locations).

Concrete blocks to BS 6073-1:1981 should have a density greater than 1500 kg/m³.

Bricks and blocks should be laid in a class (ii) mortar with a class (i) mortar being used for cappings, copings and work below DPC level. Use should be made of sulfate resisting cement if the base of the wall is likely to remain continually damp.

The tops of all walls should be protected with a capping or coping laid on either a flexible DPC of a type that had a high resistance to shear and a good flexural bond (e.g. a moulded, ribbed surface or a coarse sand facing) or on a double course of slates or creasing tiles laid 'break joint'.

Cappings and copings must be designed to shed water and remain securely in place (i.e. they must either be heavy – approx. 150kg/m² for a concrete coping to BS 5624-2:1983 – or of a type that 'locks' onto the wall). The ends of all runs of coping must be secured, either with metal cramps or by way of the use of 'special shaped' masonry units. All copings should overhang the faces of the wall by at least 45mm, except perhaps in sheltered locations where 'flush' copings may be appropriate.

A DPC is not required at low level if the wall is entirely con-structed of frost resistant masonry. Otherwise, a DPC compris-ing two courses of DPC quality bricks should be provided; *on no account should a flexible sheet or slate DPC be used at low level, since this may cause the wall to slip or overturn.*

Movement joints should be provided at 6m, 9m and 12m cen-tres and within 3m, 4.5.m and 6m of the end of any wall for clay, calcium silicate and concrete masonry respectively.

A freestanding wall (sheltered location)

COMMON MISTAKES

- Frost resistant masonry not used for cappings or where wall exposed to high levels of wetting.
- No DPC beneath copings or cappings.
- Insufficient overhang or drip to cappings and copings.
- Cappings or copings too light or not secured in place.
- Clay DPC bricks used in wall constructed of concrete or calcium silicate (possibility of differential movement).
- Flexible sheet or slate DPC used at base of wall.
- Movement joints not taken down below DPC level.

REFERENCES

BRE GBG 14, 17, 19 & 27
BRE DAS 129
BDA DG 02 & 12

Inadequate Drainage to Rear of Retaining Walls

THE PROBLEM

An earth retaining wall is designed to counter horizontal forces exerted by the ground, and as such must be able to resist a considerable amount of pressure against its rear face. However, there comes a point when it becomes more practical – and economic – to relieve the wall of this pressure rather than to simply 'beef up' its structure, the most marked example being the generally accepted necessity to keep the ground behind the wall dry and prevent any significant build-up of groundwater against its rear face. Hence the need for the design of any earth retaining wall to incorporate some form of positive drainage, thereby preventing its premature failure.

GUIDANCE

Immediately behind the wall and in lieu of the general backfill (a tightly packed but non-cohesive material free from clay or organic material) there should be at least 450mm of free-draining granular material such as a coarse aggregate, clean gravel or crushed stone.

If the general backfill contains any fine-grained material, the 'drainage layer' must be wrapped in a layer of geotextile fabric, to prevent the migration of fines that – over time – would compromise the free-flow of water.

The drainage layer must also be separated from the back of the wall – which should be finished with a bitumen membrane, two coats of a bitumen emulsion or some similar waterproof treatment – with a layer of fibreboard or a similar material (to prevent the placement and compaction of the granular material from damaging the waterproofing).

75mm diameter weepholes at one metre centres and at least 150mm above ground level should be provided at the base of the wall, either above or within the zone of the masonry damp proof course. Alternatively, a perforated drain can be laid at the base of the drainage layer, against the back of the wall and connected to the surface water drainage system for the site. A layer of concrete fill, extending from the top of the retaining wall foundation to the level of the weepholes or a 100–150mm diameter land drain should also be placed immediately behind the wall also (to prevent water draining down and saturating the foundations and their surrounding ground).

Care should also be taken to ensure that the ground that is being retained is well-drained.

Weepholes at the base of a retaining wall

COMMON MISTAKES

- No drainage at base of wall.
- Omission of geotextile filter between drainage layer and fine-grained backfill.
- Weepholes set too high above base of wall.
- No concrete to fill zone between foundation and weep holes or land drain.

REFERENCES

BRE GBG 27
BRE DAS 130
BDA DG 02

Failure to Adequately Consider Foundations to Pavements

THE PROBLEM   RATING **3 – 4**

Very rarely is the natural soil able to support a pavement (i.e. the layers that make up a vehicular or pedestrian surface) without some form of preparation. At the very least the formation level (also known as the 'grade' – the soil below is the 'subgrade') will require shaping and compaction; and if the soil is weak, replacing or strengthening with layers of imported material. An inadequate foundation will invariably result in the rapid deterioration of the pavement, so it is essential that designers have at least a basic understanding of how paving systems are influenced by the underlying ground and how the design of their foundations should be approached.

GUIDANCE

The strength of the subgrade is described by an index known as the California Bearing Ratio (CBR), ascertained insitu or in a laboratory, or by reference to published tables. Typical CBR values are 1.5% to 3.5% for heavy clay, 2.5% to 8% for silty or sandy clay, 20% to 40% for sand, and 60% for well compacted gravel. Higher values – indicating a good bearing capacity – occur where the water table is low, the exposed soil can be promptly protected from wetting, and the site is well drained; if in doubt, the worst conditions should be assumed.

Having established a CBR for the subgrade, it can then be decided if an 'improvement' (capping) layer of granular or treated material is required, and if so how thick it should be. For lightly trafficked pavements (i.e. roads with less than 30 commercial vehicles a day and a 30mph speed limit, parking and pedestrian areas), a capping layer 350mm thick will be necessary if the CBR is less than 5% (600mm if 2% or less, though specialist engineering advice should be sought if below 1%). Exceptionally soft areas of subgrade (e.g. peat) should be removed and replaced with properly compacted and inert material of a similar strength to the surrounding ground.

The subgrade must be overlain with a sub-base of compacted granular or cement-bound material (or lean mix concrete) at least 150mm thick (225mm if there is no capping layer), thicker if it is to form the temporary surface of a site access road.

If the subgrade is hard rock or granular material with a laboratory CBR in excess of 30%, the sub-base may be omitted. Sub-base material may also be used in lieu of a capping layer where the CBR is above 2.5%, allowing a reduced thickness of foundation (sub-base material is of a higher quality than that used for the cheaper capping). Reduced thicknesses *may* be suitable for very lightly trafficked surfaces, though 150mm is often the minimum thickness that can be spread and compacted. Where the soil is susceptible to frost, the thickness should be increased to ensure that the pavement has at least 450mm of non frost-susceptible material. The surface of the sub-base should be close-textured; blinding or a geotextile layer may be required.

Special conditions may apply to pavements that are to be adopted by the local authority.

Concrete paving with no foundation

COMMON MISTAKES

- Failure to base design of pavement on assessed CBR.

- Soft spots or especially weak material not replaced.

- Inadequate depth of capping and sub-base to lightly loaded pavements (as much as 0.5m+ needed for a CBR of 2% or lower).

- Thickness of sub–base not increased to account for use by construction traffic.

- No preparation of sub–base for laying of roadbase or bedding course.

REFERENCES

BS 6677–2:1986 cl.4

BS 7533:1992 cl.4

BS 7533–3:1997 cl.4

BS 7533–4:1998 cl.4

BCA 46.030

BDA DG 21

BDA DN 08 & DN 09

BDA SP 15 & SP 16

HA DMRB7:HD 25/94

Interpave 1998

Inappropriate or Poorly Bedded Surfacing to Pavements

THE PROBLEM   RATING 3 – 4

The performance of a vehicular or pedestrian pavement is heavily dependent on the careful specification and detailing of both the surfacing material and its bedding. Loading (vehicular or pedestrian), wear and abrasion, exposure to the weather and the control of movement must all be taken into account if the displacement, cracking and premature break up of pavements is to be avoided.

GUIDANCE

Lightly trafficked roads, and pedestrian pavements do not require a roadbase (a structural layer beneath the surfacing), except perhaps where vehicular traffic is 'channelled' or the base or subgrade are moisture or frost sensitive.

Paving blocks (pavers) are laid on a 50mm thick bedding of well compacted sand (35mm if laid on a bound roadbase) with all joints filled with free-flowing, dry sand. The pavement must be compacted using a plate vibrator, the joints being refilled after each pass. Concrete pavers should comply with BS 6717–1:1993, clay and calcium silicate to BS 6677–1:1986. Type PA clay pavers are only for pedestrian areas, car parks and driveways. Calcium silicate pavers are unsuited to repeated de-icing with salts or wetting with sea water. Pavers to roads should have chamfered edges and – if rectangular – be laid in a 'herringbone' pattern.

Flags or slabs up to 450mm x 450mm are laid on a bed of compacted sand (50mm if absolutely no vehicular access, 30mm otherwise); larger units must be laid on a 25mm bed of mortar (1:3 cement or lime-sand). Wide joints (5–10mm on mortar bedding only) should be filled with mortar, narrow joints with sand (the paving must be compacted with a plate vibrator). Precast concrete flags should comply with BS 7263–1:1984. Flags that might be subject to vehicular loading should be no larger than 450mm x 450mm and at least 60mm thick, or 600mm x 600mm x 63mm if only used as crossing serving a house.

Asphalt and Macadam is usually in two layers, typically a 50mm base course with a 20–30mm wearing course to pedestrian areas, 40–50mm to vehicular. Systems should be to BS 594 (hot rolled asphalt), BS 1466 & 1477 (mastic asphalt) or BS 4987 (macadam).

Insitu concrete paving should be reinforced, at least 100mm thick and laid in bays with movement joints at 4m centres. Provision for movement should also be made at walls, and bays should be joined with steel dowels (to restrain vertical movement). Although combining the functions of roadbase and surface, the rigidity of insitu concrete means that its use might not be appropriate where the subgrade has significant potential for movement (e.g. heavy, shrinkable clay), even if laid on a substantial foundation.

Pavements must be effectively drained and all types – except rigid insitu concrete – restrained at the edges (e.g. by a kerb).

Pavers laid in herringbone pattern

COMMON MISTAKES

- Rectangular pavers to roads laid in a 'running bond' or 'basket weave' pattern.

- Wrong thickness of bedding to mixed paving (thickness should be dictated by flags).

- No edge restraint to flexible pavements (i.e. pavers, flags, asphalt or macadam).

- Insufficient movement joints or no dowels between bays of insitu concrete paving.

REFERENCES

BS 6677–2:1986 cls.3 & 5–8
BS 7533:1992 cls.4.3.5–7 & 5
BS 7533–3:1997 cls.4.3 & 4.4
BS 7533–4:1998 cl.6
BCA 46.030
BDA DG 21
BDA DN 08 & DN 09
BDA SP 15 & SP 16
Interpave 1998
MAC 1999

References

BRITISH STANDARDS INSTITUTION (BSI)

British Standards (BS)

BS 187:1978 *Specification for calcium silicate (sandlime and flintlime) bricks*

BS 493:1995 *Specification for airbricks and gratings for wall ventilation*

BS 585–2:1985 *Wood stairs. Specification for performance requirements for domestic stairs constructed of wood-based materials*

BS 594–1:1992 *Hot rolled asphalt for roads and other paved areas. Specification for constituent materials and asphalt mixes*

BS 594–2:1992 *Hot rolled asphalt for roads and other paved areas. Specification for transport, laying and compaction of rolled asphalt*

BS 747:1994 *Specification for roofing felts*

BS 1217:1997 *Specification for cast stone*

BS 1243:1978 *Specification for metal ties for cavity wall construction*

BS 1297:1987 *Specification for tongued and grooved softwood flooring*

BS 1466:1973 *Specification for mastic asphalt (natural rock asphalt fine aggregate) for roads and footways*

BS 1477:1988 *Mastic asphalt (limestone fine aggregate) for roads, footways and paving in building*

BS 1521:1972 *Specification for waterproof building papers*

BS 3921:1985 *Specification for clay bricks*

BS 4483:1998 *Steel fabric for the reinforcement of concrete*

BS 4987–1:1993 *Coated macadam for roads and other paved areas. Specification for constituent materials and mixtures*

BS 4987–2:1993 *Coated macadam for roads and other paved areas. Specification for transport, laying and compaction*

BS 5234–1:1992 *Partitions(including matching linings). Code of practice for design and installation*

BS 5250:1989 *Code of practice for control of condensation in buildings*

BS 5262:1991 *Code of practice for external renderings*

BS 5268–2:1991 *Structural use of timber. Code of practice for permissible stress design, materials and workmanship*

BS 5268–3:1985 *Structural use of timber. Code of practice for trussed rafter roofs*

BS 5268–6.1:1996 *Structural use of timber. Code of practice for timber frame walls: Dwellings not exceeding four storeys*

BS 5390:1976 *Code of practice for stone masonry*

BS 5410–1:1997 *Code of practice for oil firing. Installations up to 45kW output capacity for space heating and hot water supply purposes*

BS 5440–1:1990 *Installation of flues and ventilation for gas appliances of rated input not exceeding 60kW. Specification for installation of flues*

BS 5440–2:2000 *Installation of flues and ventilation for gas appliances of rated input not exceeding 60kW. Specification for installation of ventilation for gas appliances*

BS 5449:1990 *Specification for forced circulation hot water central heating systems for domestic premises*

BS 5534–1:1997 *Code of practice for slating and tiling. Design*

BS 5572:1994	*Code of practice for sanitary pipework*
BS 5618:1996	*Code of practice for thermal insulation of cavity walls (with masonry or concrete inner and outer leaves) by filling with urea-formaldehyde (UF) foam systems*
BS 5628–1:1992	*Code of practice for the use of masonry. Structural use of unreinforced masonry*
BS 5628–3:1985	*Code of practice for the use of masonry.* *Materials and components, design and workmanship*
BS 5642–1:1978	*Sills and copings.* *Sills of precast concrete, cast stone, clayware, slate and natural stone*
BS 5642–2:1983	*Sills and copings.* *Copings of precast concrete, cast stone, clayware, slate and natural stone*
BS 5713:1979	*Specification for hermetically sealed flat double glazing units*
BS 5720:1979	*Code of practice for mechanical ventilation and air conditioning in buildings*
BS 5839–6:1995	*Fire detection and alarm systems for buildings. Code of practice for the design and installation of fire detection and alarm systems in dwellings*
BS 5871–1:1991	*Specification for installation of gas fires, convector heaters, fire/back boilers and decorative fuel effect gas appliances.* *Gas fires, convector heaters and fire/back boilers (1st, 2nd and 3rd family gases)*
BS 5871–2:1991	*Specification for installation of gas fires, convector heaters, fire/back boilers and decorative fuel effect gas appliances.* *Inset live fuel effect gas fires of heat input not exceeding 15kW (2nd & 3rd family gases)*
BS 5871–3:1991	*Specification for installation of gas fires, convector heaters, fire/back boilers and decorative fuel effect gas appliances. Decorative fuel effect gas appliances of heat input not exceeding 15 kW (2nd and 3rd family gases)*
BS 5925:1991	*Code of practice for ventilation principles and designing for natural ventilation*
BS 5930:1981	*Code of practice for site investigations*
BS 5977–2:1983	*Lintels. Specification for prefabricated lintels*
BS 6073–1:1981	*Precast masonry units. Specification for precast concrete masonry units*
BS 6178–1:1990	*Joist hangers.* *Specification for joist hangers for building into masonry walls of domestic dwellings*
BS 6229:1982	*Code of practice for flat roofs with continuously supported coverings*
BS 6262:1982	*Code or practice for glazing buildings.*
BS 6297:1983	*Code of practice for design and installation of small sewage treatment works and cesspools*
BS 6367:1983	*Code of practice for drainage of roofs and paved areas*
BS 6457:1984	*Specification for reconstructed stone masonry units*
BS 6676–2:1986	*Thermal insulation of cavity walls using man-made mineral fibre batts (slabs).* *Code of practice for installation of batts (slabs) filling the cavity*
BS 6677–1:1986	*Clay and calcium silicate pavers for flexible pavements. Specification for pavers*
BS 6677–2:1986	*Clay and calcium silicate pavers for flexible pavements.* *Code of practice for design of lightly trafficked pavements*
BS 6700:1997	*Specification for design, installation, testing and maintenance of services supplying water for domestic use within buildings and their curtilages*
BS 6717–1:1993	*Precast concrete paving blocks. Specification for paving blocks*

BS 6798:1987	*Specification for installation of gas-fired hot water boilers of rated input not exceeding 60kW*
BS 6915:1988	*Specification for design and construction of fully supported lead sheet roof and wall coverings*
BS 6925:1988	*Specification for mastic asphalt for building and civil engineering (limestone aggregate)*
BS 7263-1:1994	*Precast concrete flags, kerbs, channels, edgings and quadrants. Specification*
BS 7430:1991	*Code of practice for earthing*
BS 7533:1992	*Structural design of pavements constructed with clay or concrete block pavers*
BS 7533-3:1997	*Pavements constructed with clay, natural stone or concrete pavers. Code of practice for laying precast concrete paving blocks and clay paves for flexible pavements*
BS 7533-4:1998	*Pavements constructed with clay, natural stone or concrete pavers. Code of practice for the construction of pavements of precast concrete flags or natural stone slabs*
BS 7619:1993	*Specification for extruded cellular unplasticized PVC (PVC-UE) profiles*
BS 7671:1992	*Requirements for electrical installations. IEE Wiring Regulations. Sixteenth edition.*
BS 7916:1998	*Code of practice for the selection and application of particleboard, oriented strand board (OSB), cement bonded particleboard and wood fibreboards for specific purposes*
BS 8000-5:1990	*Workmanship on building sites. Code of practice for carpentry, joinery and general fixings*
BS 8000-8:1994	*Workmanship on building sites. Code of practice for plasterboard partitions and dry linings*
BS 8004:1986	*Code of practice for foundations*
BS 8102:1990	*Code of practice for protection of structures against water from the ground*
BS 8103-1:1995	*Structural design of low-rise buildings. Code of practice for stability, site investigations, foundations and ground floor slabs for housing*
BS 8103-2:1996	*Structural design of low-rise buildings. Code of practice for masonry walls for housing*
BS 8103-3:1996	*Structural design of low-rise buildings. Code of practice for timber floors and roofs for housing*
BS 8103-4:1995	*Structural design of low-rise buildings. Code of practice for suspended concrete floors for housing*
BS 8104:1992	*Code of practice for assessing exposure of walls to wind-driven rain*
BS 8201:1987	*Code of practice for flooring of timber, timber products and wood based panel products*
BS 8204-1:1987	*Screeds, bases and in-situ floorings. Code of practice for concrete bases and screeds to receive in-situ flooring*
BS 8212:1995	*Code of practice for dry lining and partitioning using gypsum wallboard*
BS 8215:1991	*Code of practice for design and installation of damp-proof courses in both solid and cavity masonry construction*
BS 8217:1994	*Code of practice for built-up felt roofing*
BS 8218:1998	*Code of practice for mastic asphalt roofing*
BS 8233:1999	*Sound insulation and noise reduction for buildings. Code of practice*
BS 8301:1985	*Code of practice for building drainage*

Codes of Practice (BS CP)

BS CP 102:1973   *Code of practice for protection of buildings against water from the ground (Partially replaced)*

BS CP 143–5:1964   *Code of practice for sheet roof and wall coverings. Zinc*

BS CP 143–12:1970   *Code of practice for sheet roof and wall coverings. Copper*

Drafts for Development (BS DD)

BS DD 140–2:1987   *Wall ties. Recommendations for design of wall ties*

BS DD 175:1988   *Code of practice for the identification of potentially contaminated land and its investigation*

European Standards adopted as British Standards (BS EN)

BS EN 300:1997   *Oriented strand boards (OSB). Definitions, classifications and specifications*

BS EN 309:1992   *Wood particleboards. Definition and classification*

BS EN 501:1994   *Roofing products from metal sheet – Specification for fully supported roofing products of zinc sheet*

BS EN 622–5:1997   *Fibreboards, Specifications. Requirements for dry process boards (MDF)*

BS EN 633:1994   *Cement-bonded particleboards. Definition and classification*

BS EN 636–1:1997   *Plywood. Specifications. Requirements for plywood for use in dry conditions*

BS EN 636–2:1997   *Plywood. Specifications. Requirements for plywood for use in humid conditions*

BS EN 752–3:1996   *Drain and sewer systems outside buildings. Planning.*

BS EN 942:1996   *Timber in joinery. General classification of timber quality*

BS EN 1313–1:1997   *Round and sawn timber. Permitted deviations and preferred sizes. Softwood sawn timber*

BUILDING RESEARCH ESTABLISHMENT (BRE)

Defects Action Sheets (DAS)

18   *External masonry walls: vertical joints for thermal and moisture movement* (1983)

25   *External and separating walls: lateral restraint at intermediate floors – specification* (1983)

31   *Suspended timber floors: chipboard flooring – specification* (1983)

35   *Substructure: DPCs and DPMs – specification* (1983)

36   *Substructure: DPCs and DPMs – installation* (1983)

41   *Plastics sanitary pipework: jointing and support – specification* (1983)

42   *Plastics sanitary pipework: jointing and support – installation* (1983)

55   *Roofs: eaves gutters and downpipes – specification* (1984)

61   *Cold water storage cisterns: overflow pipes* (1985)

62   *Electrical services: avoiding cable overheating* (1985)

68   *External walls: joints with windows and doors – detailing for sealants* (1985)

70   *External masonry walls: eroding mortars – repoint or rebuild?* (1986)

71   *External masonry walls: repointing – specification* (1986)

72   *External masonry walls: repointing* (1986)

73   *Suspended timber ground floor: remedying dampness due to inadequate ventilation* (1986)

75	*External walls: brick cladding to timber frame: the need to design for differential movement* (1986)
81	*Plasterboard ceilings for direct decoration: nogging and fixing – specification* (1986)
92	*Balanced flue terminals: location and guarding* (1987)
96	*Foundations on shrinkable clay: avoiding damage due to trees* (1987)
98	*Wood windows: resisting rain penetration at perimeter joints* (1987)
99	*Suspended timber floors: notching and drilling of joists* (1987)
101	*Plastics sanitary pipework: specifying for outdoor use* (1987)
106	*Cavity parapets-avoiding rain penetration* (1987)
109	*Hot and cold water systems-protection against frost* (1987)
114	*Slated and tiled pitched roofs: flashings and cavity trays for step and stagger layouts* (1988)
115	*External masonry cavity walls: wall ties – selection and specification* (1988)
142	*Slate clad roofs: fixing of slates and battens* (1990)
143	*Drainage stacks: avoiding roof penetration* (1990)

Digests

54	*Damp proofing solid floors* (1971)
63	*Soils and foundations. Part 1* (1980)
64	*Soils and foundations. Part 2* (1972)
67	*Soils and foundations. Part 3* (1980)
157	*Calcium silicate (sandlime, flintlime) brickwork* (1992)
170	*Ventilation of internal bathrooms and WCs in dwellings* (1981)
240	*Low-rise buildings on shrinkable clay soils. Part 1* (1993)
241	*Low-rise buildings on shrinkable clay soils. Part 2* (1990)
242	*Low-rise buildings on shrinkable clay soils. Part 3* (1980)
276	*Hardcore* (1983)
298	*The influence of trees on house foundations in clay soils* (1985)
306	*Domestic draughtproofing: ventilation considerations* (1986)
311	*Wind scour of gravel ballast on roofs* (1986)
312	*Flat roof design: the technical options* (1988)
323	*Selecting wood-based panel products* (1992)
324	*Flat roof design: thermal insulation* (1987)
333	*Sound insulation of separating walls and floors: Part 1: walls* (1998)
334	*Sound insulation of separating walls and floors: Part 2: floors* (1989)
348	*Site investigation for low-rise building: the walk-over survey* (1989)
361	*Why do buildings crack?* (1991)
362	*Building mortar* (1991)
363	*Sulfate and acid resistance of concrete in the ground* (1996)
364	*Design of timber floors to prevent decay* (1991)
365	*Soakaway design* (1991)
372	*Flat roof design: water proof membranes* (1992)
380	*Damp-proof courses* (1993)
395	*Slurry trench cut-off walls to contain contamination* (1994)

398	*Continuous mechanical ventilation in dwellings: design, installation and operation* (1994)
410	*Cementitious renders for external walls* (1995)
419	*Flat roof design: bituminous waterproof membranes* (1996)
420	*Selecting natural building stones* (1997)
427–1	*Low-rise buildings on fill: Classification and load-carrying characteristics* (1997)
427–2	*Low-rise buildings on fill: Site investigation, ground movement and foundation design* (1997)
427–3	*Low-rise buildings on fill: Engineered fill* (1997)
435	*Medium density fibreboard* (1998)
441	*Clay bricks and masonry* (1999)

Good Building Guides (GBG)

8	*Bracing trussed rafter roofs* (1991)
14	*Building simple plan brick or blockwork freestanding walls* (1994)
16	*Erecting, fixing and strapping trussed rafter roofs* (1993)
17	*Freestanding brick walls – repairs to copings and cappings* (1993)
18	*Choosing external rendering* (1994)
19	*Building reinforced, diaphragm and wide plan freestanding walls* (1994)
21	*Joist hangers* (1996)
25	*Buildings and radon* (1996)
27	*Building brickwork or blockwork retaining walls* (1996)
28–1	*Domestic floors: construction, insulation and damp-proofing* (1997)
29–1	*Connecting walls and floors. A practical guide* (1997)
30	*Carbon monoxide detectors* (1999)
33	*Building damp free cavity walls* (1999)
35	*Building without cold spots* (1999)
36	*Building a new felted flat roof* (1999)
37	*Insulating roofs at rafter level: sarking insulation* (2000)
38	*Disposing of rainwater* (2000)

Information Papers (IP)

04/81	*Performance of cavity wall ties* (1981)
06/86	*Spacing of wall ties in cavity walls* (1986)
02/87	*Fire and explosion hazards associated with the redevelopment of contaminated land* (1987)
16/88	*Ties for cavity walls: new developments* (1988)
17/88	*Ties for masonry cladding* (1988)
18/88	*Domestic mechanical ventilation: guidelines for designers and installers* (1988)
05/89	*Use of 'vibro' ground improvement techniques in the United Kingdom* (1989)
21/89	*Passive stack ventilation in dwellings* (1989)
08/91	*Mastic asphalt for flat roofs: testing for quality assurance* (1991)
10/93	*Avoiding latent mortar defects in masonry* (1993)
07/94	*Spillage of flue gasses from solid-fuel combustion appliances* (1994)
12/94	*Assessing condensation risk and heat loss at thermal bridges round openings* (1994)

13/94 *Passive stack ventilation systems: design and installation* (1994)

06/95 *Flow resistance and wind performance of some common ventilation terminals* (1995)

07/95 *Bituminous roofing membranes: performance in use* (1995)

Reports

29 *The building limestones of France* (1982)

36 *The building limestones of the British Isles* (1983)

84 *The building sandstones of the British Isles* (1986)

134 *The building magnesian limestones of the British Isles* (1988)

162 *Background ventilation of dwellings: a review* (1989)

211 *Radon: guidance on protective methods for new dwellings* (1999)

212 *Construction of new buildings on gas-contaminated land* (1991)

232 *Recognising wood rot and insect damage in buildings* (1992)

233 *Briefing guide for timber-framed housing* (1993)

238 *Sound control for homes* (1993)

262 *Thermal insulation avoiding risks* (1994)

280 *Double-glazing units: a BRE guide to improved durability* (1995)

302 *Roofs and roofing* (1996)

332 *Floors and flooring* (1997)

352 *Walls, windows and doors* (1998)

358 *Quiet homes: a guide to good practice and reducing the risk of poor sound insulation between dwellings* (1998)

376 *Radon: Guidance on protective measures for new dwellings in Scotland* (1999)

391 *Specifying vibro stone columns* (2000)

BRICK DEVELOPMENT ASSOCIATION (BDA)

Design Guides (DG)

02 *The Design of Brickwork Retaining Walls* (1991)

12 *Design of Freestanding Walls* (1984)

21 *The Design of Flexible Pavements Surfaced with Clay Pavers* (1990)

Design Notes (DN)

08 *Rigid Paving with Clay Pavers* (1995)

09 *Flexible Paving with Clay Pavers* (1998)

10 *Designing for Movement in Brickwork* (1986)

16 *Resisting Rain Penetration with facing brickwork* (1997)

Special Publications (SP)

15 *Specification for Clay Pavers for Flexible Pavements* (1988)

16 *Code of Practice for Flexible Pavements Constructed in Clay Pavers* (1988)

Good Practice Note (GPN)

CIW-2 *Cavity insulated walls specifiers guide* (1987)

BRITISH CEMENT ASSOCIATION (BCA)

46.030   *Concrete Pavements for Highways* (1992)

BRITISH GAS (BG)

DM3   *Specification for flues for class II appliances in timber framed houses* (1984)

BRITISH PLASTICS FEDERATION (BPF)

348/2   *Code of practice for the Installation of PCV-U Windows and Doorsets* (1996)

349/1   *Code of practice for the Installation of cellular PVC-U Cladding Systems* (1994)

356/1   *Code of practice for the Installation of PVC-U Windows and Doorsets*
                    in New Domestic Dwellings (1997)

BUILDING SERVICES RESEARCH AND INFORMATION ASSOCIATION (BSRIA)

Application Guides/Handbooks

87/1   *Operating and maintenance manuals for building services installation* (1990)

89/2   *Commissioning of water systems in buildings* (1989)

92/3   *Installation, commissioning and maintenance of fire and security systems* (1992)

CLAY PIPE DEVELOPMENT ASSOCIATION (CPDA)

Technical Notes (TN)

01   *Laying Vitrified Pipes in Soft Ground* (1995)

02   *Laying Vitrified Pipes in Waterlogged Ground* (1995)

04   *Concrete Bedding to Vitrified Clay Pipelines* (1995)

06   *Laying Vitrified Pipes at Shallow Depths* (1995)

Other Publications

                    The specification, design and construction of drainage and sewerage systems
                    using vitrified clay pipes (1999)

                    The use of clay flue liners and terminals (1996)

CONSTRUCTION INDUSTRY RESEARCH AND INFORMATION ASSOCIATION (CIRIA)

Reports

127   *Sound control for homes* (1993)

130   *Methane: Its occurrence and hazards in construction* (1993)

131   *The measurement of methane and other gases from the ground* (1993)

149   *Methane and associated hazards to construction:*
                    Protecting development from methane (1995)

Special Publications (SP)

124   *Barriers, liners and cover systems for containment and control of land contamination* (1996)

144/L2   *On-site sewage dispersal options* (1998)

Technical Notes (TN)

146	*Septic tanks and small sewage treatment works* (1993)

COPPER DEVELOPMENT ASSOCIATION (CDA)

Technical Notes (TN)

32	*Copper in roofing: Design and installation* (1995)
33	*Copper tube in domestic water services* (1998)
39	*Copper in domestic heating systems* (1998)

FIBRE BUILDING BOARD ORGANISATION (FIDOR)
[A predecessor to the Wood Panel Industries Federation]

X509	*Use of MDF in the manufacture of stairs-general guidance* (1988)
SR6	*Fibre building boards technical information* (1991)

GLASS AND GLAZING FEDERATION (GGF)

Glazing Manual Data Sheets (GM)

4.2	*System design and glazing considerations for insulating glass units* (1995)
4.2.1	*Insulating glass units – guidance notes for the specification and selection of various types* (1991)
5.16	*Glazing techniques for timber windows with microporous stain finishes* (1989)

HIGHWAYS AGENCY (HA)

Design Manual for Roads and Bridges (DMRB)

Volume 7 (DMRB7)	*Pavement Design and Maintenance:–*
HD 23/99	*Section 1 Part 2 General Information* (1999)
HD 25/94	*Section 2 Part 2 Foundations* (1994)
HD 26/94	*Section 2 Part 3 Pavement Design* (1998)

HOUSING ASSOCIATION PROPERTY MUTUAL (HAPM)

Technical Notes (TN)

02	*Exposure ratings and weather exclusion* (1994)
05	*Timber noggings and plasterboard lining* (1995)
08	*Sound insulation in new dwellings* (1997)
09	*Safe access into houses for disabled people* (1997)
11	*Particleboard Floor Decking* (1997)
12	*Insulating glass units in timber windows* (1999)

INSTITUTE OF PLUMBING (IOP)

Plumbing engineering services design guide (PESDG)

Section A	*Hot and Cold Water Supplies* (1988)
Section B	*Domestic Central Heating* (1988)
Section C	*Sanitary Plumbing and Drainage* (1988)
Section M	*Electrical Earthing and Bonding in Water Installations* (1988)

INTERDEPARTMENTAL COMMITTEE FOR THE REDEVELOPMENT OF CONTAMINATED LAND

ICRCL Note 59/83 (1987)

INTERPAVE (THE PRECAST CONCRETE PAVING AND KERB ASSOCIATION)

Design Handbook (1998)

LEAD SHEET ASSOCIATION (LSA)

The Lead Sheet Manual. Volume 1: *Lead Sheet Flashings* (1990)

The Lead Sheet Manual. Volume 2: *Lead Sheet Roofing and Cladding* (1992)

The Lead Sheet Manual. Update 1: *Pitched valley gutters* (1991)

The Lead Sheet Manual. Update 2: *Underside Corrosion* (1993)

MASTIC ASPHALT COUNCIL (MAC)

Mastic Asphalt – The Technical Guide (1999)

NATIONAL HOUSE BUILDING COUNCIL (NHBC)

NBHC Standards

Chapter 4.1	*Land quality-managing ground conditions* (1999)
Chapter 4.2	*Building near trees* (1999)
Chapter 4.4	*Strip and trench fill foundations* (1999)
Chapter 4.5	*Raft, pile, pier and beam foundations* (1999)
Chapter 4.6	*Vibratory ground improvement techniques* (1999)
Chapter 5.1	*Substructure and ground bearing floors* (1999)
Chapter 5.2	*Suspended ground floors* (1999)

NATIONAL RADIOLOGICAL PROTECTION BOARD (NRPB)

Reports

254	*Radon in dwellings in England* (1992)
272	*Exposure to radon in UK dwellings* (1994)
290	*Radon Atlas of England* (1996)

NATIONAL STANDARDS AUTHORITY OF IRELAND (NSAI)

Standard Recommendations (SR)

SR6:*1991*   *Septic tank systems. Recommendations for domestic effluent treatment and disposal from a single family dwelling house*

PACKAGING AND INDUSTRIAL FILMS ASSOCIATION (PIFA)

Standard 6/83A (1995)

STEEL WINDOW ASSOCIATION (SWA)

Fact Sheets (FS)

03   *Fixing* (1990)

TIMBER RESEACH AND DEVELOPMENT ASSOCIATION (TRADA)

Wood Information Sheets

0–03   *Introduction to timber framed housing* (1992)

1–10   *Principles of pitched roof construction* (1993)

1–20   *External timber cladding* (1993)

1–29   *Trussed rafters* (1991)

1–36   *Timber joist and deck floors – avoiding movement* (1995)

4–27   *Moisture content standards for timber* (1996)

Other Publications

Timber Frame Construction (1994)

Wood windows – Design, selection and installation (1993)

TRUSSED RAFTER ASSOCIATION (TRA)

SFG G12   *Technical Handbook & Site Installation Guide* (1999)

WOOD PANEL INDUSTRIES FEDERATION (WPIF)

Code of practice for particleboard and oriented strand board (OSB) floating floors (1998)

ZINC DEVELOPMENT ASSOCIATION (ZDA)

Zinc in Building Design (1988)

Index